Bernard Perez

**The first three years of childhood**

Bernard Perez

**The first three years of childhood**

ISBN/EAN: 9783337215316

Printed in Europe, USA, Canada, Australia, Japan

Cover: Foto ©berggeist007 / pixelio.de

More available books at **www.hansebooks.com**

# THE FIRST THREE YEARS
## OF CHILDHOOD

BY

BERNARD PEREZ

EDITED AND TRANSLATED BY
## ALICE M. CHRISTIE
*Translator of "Child and Child Nature," etc., etc.*

WITH AN INTRODUCTION BY JAMES SULLY, M.A.
*Author of "Outlines of Psychology," etc.*

STEREOTYPED     EDITION.

London:
### SWAN SONNENSCHEIN & CO.,
PATERNOSTER SQUARE,

1892.

# PREFACE.

FOUR years ago this book appeared to be a fortunate hit, offering as it did, though on the most simple scale, a study of infant psychology. The character of the work arose naturally from the line of study I had planned. I had, in fact, set myself to follow out in little children the gradual awakening of those faculties which constitute the psychic activity, so abundantly differentiated, so delicate and at the same time so powerful, of the adult human being. It was, no doubt, this intention—possibly a somewhat premature one—of systematizing a class of observations, of which hitherto only rough sketches had been attempted, which gained me the encouragement of philosophers and educationalists, both in France and abroad. Thus then, while endeavouring to improve upon the modest beginning which had at first won me their sympathy, I could do no better than keep to my original method. My canvases were already sketched in and accepted: I have merely endeavoured to draw them a little better and to fill them in more,

to render facts and interpretations of facts clearer and
more precise.

With regard to the facts, either simply described or
dramatized in the form of psychological anecdotes, I
have carefully sorted and re-arranged them, discard-
ing a certain number which seemed to me of little
importance, and adding a good many others, either
taken out of my own journals of observations or bor-
rowed from other people, but all of them verified
by myself. I thought also that I should be readily
forgiven for having interpolated a few pages of psychological observations taken out of my book on
*Education from the Cradle*, which seemed here in
their natural place, and which will be replaced in the
other book by considerations of a more specially peda-
gogic nature. Both books will, I think, have gained
by the exchange.

As to the interpretations of facts, I have striven to
be guided by the spirit of the experimental method.
If I have sometimes been happy in my observations
and judgments, it is to this method that the honour
is due; the mistakes and omissions are my own
share. At any rate, no one of the systems of philosophy, which, under different names, have more or
less exactly adapted themselves to the experimental
method, is responsible for my errors. Although I
have my preferences and my tendencies, I belong to
no school. I find myself most often, it is true, quot-

ing from a Darwin or a Spencer, but I am none the
less glad to do so from Mme. Necker, de Saussure,
and Guizot, from Messrs. E. Egger and L. Ferri,
when they can give me the fruits of real experience.
I do not ask of facts and ideas for their label and
trade-mark before admitting them to my humble psychological domain ; it is enough for me that they are
facts well observed and well described, enough that
they are clear and judicious ideas.

It is in such a spirit that I would have my readers
deal with my essay on infant psychology ; letting all
thoughts of a particular system be secondary in their
minds, as they have been in my own. I have often
pondered—trying to turn them to account myself—
those most pithy words of Mr. F. Pollock's : "And as
science makes it plainer every day that there is no
such thing as a fixed equilibrium, either in the world
without or in the mind within, so it becomes plain
that the genuine and durable triumphs of philosophy
are not in systems but in ideas."[1]

If some few good ideas are found scattered in my
book, I beg my readers neither to see or seek for
anything else in it.

BERNARD PEREZ.

---

[1] *Spinoza : His Life and Philosophy*, p. 408

# CONTENTS.

—✦—

## CHAPTER I.

|       |                                                      | PAGE |
|-------|------------------------------------------------------|------|
| I.    | THE FACULTIES OF THE INFANT BEFORE BIRTH.            | 1    |
| II.   | THE FIRST IMPRESSIONS OF THE NEW-BORN CHILD         | 7    |

## CHAPTER II.

| I.    | MOTOR ACTIVITY AT THE BEGINNING OF LIFE             | 11   |
| II.   | MOTOR ACTIVITY AT SIX MONTHS                        | 16   |
| III.  | MOTOR ACTIVITY AT FIFTEEN MONTHS                    | 20   |

## CHAPTER III.

| I.    | INSTRUCTIVE AND EMOTIONAL SENSATIONS               | 23   |
| II.   | THE FIRST PERCEPTIONS                               | 32   |

## CHAPTER IV.

| I.    | GENERAL AND SPECIAL INSTINCTS                       | 44   |
| II.   | SPECIAL INSTINCTS.                                   | 49   |

## CHAPTER V.

| THE SENTIMENTS | | 60 |

## CHAPTER VI.

| I.    | INTELLECTUAL TENDENCIES                             | 82   |
| II.   | VERACITY                                            | 86   |
| III.  | IMITATION                                           | 89   |
| IV.   | CREDULITY                                           | 94   |

## CHAPTER VII.

| THE WILL | | 99 |

## CHAPTER VIII.

PAGE

THE FACULTIES OF INTELLECTUAL ACQUISITION AND RETEN-
TION.

I. ATTENTION . . . . . . . . . 110
II. MEMORY . . . . . . . . . 121

## CHAPTER IX.

THE ASSOCIATION OF PSYCHICAL STATES.

I. ASSOCIATION . . . . . . . . 131
II. IMAGINATION . . . . . . . . 147
III. SPECIAL IMAGINATION . . . . . . 152

## CHAPTER X.

ON THE ELABORATION OF IDEAS.

I. JUDGMENT . . . . . . . . 164
II. ABSTRACTION . . . . . . . . 178
III. COMPARISON . . . . . . . . 190
IV. GENERALIZATION . . . . . . . 197
V. REASONING . . . . . . . . 210
VI. THE ERRORS AND ILLUSIONS OF CHILDREN . . . 224
VII. ERRORS OWING TO MORAL CAUSES . . . 233

## CHAPTER XI.

ON EXPRESSION AND LANGUAGE, PARTS I., II., III. 236, 241, 251

## CHAPTER XII.

THE ÆSTHETIC SENSE IN LITTLE CHILDREN.

I. THE MUSICAL SENSE . . . . . . 265
II. THE SENSE OF MATERIAL BEAUTY . . . 270
III. THE CONSTRUCTIVE INSTINCT . . . . 276
IV. THE DRAMATIC INSTINCT . . . . . 278

## CHAPTER XIII.

I. PERSONALITY.—REFLECTION.—MORAL SENSE . . 282
II. THE MORAL SENSE . . . . . . 287

# INTRODUCTION.

AMONG the many new fields of investigation which modern science has opened up, there is none which is more inviting than that of infant psychology. The beginnings of all things are full of interest, as we see by the amount of inquiry now devoted to the origin of human institutions and ideas, and all the various forms of life. And the beginnings of a human mind, the first dim stages in the development of man's God-like reason, ought surely to be most interesting of all. And infancy has its own peculiar charm. There is an exquisite poetry in the spontaneous promptings of the unsophisticated spirit of the child. So far removed at times from our one-sided prejudiced views, so high above our low conventional standards are the little one's intuitions of his new world. Childhood has its unlovely and unworthy side no doubt. Still I cannot think that any close observer of infancy ever thoroughly believed in its total depravity. Possibly, indeed, to a perfectly candid mind its fresh and striking

observations about things, which, though often
bizarre, are on the whole thoroughly sound
and wholesome, are always apt to suggest the
pleasing fancy of Plato and Wordsworth, that
the little new-comer brings from his ante-natal
abode ideas and feelings which lie high above
the plane of earthly experience.   However this
be, no thoroughly open and unspoiled mind can
fail to learn much that is good from a close
study of childhood.   This is the period when
very ordinary mortals display something remarkable.   Perhaps indeed no healthy child has
ever failed to present some new mental or
moral phenomenon, to impress, amuse, or instruct, if only the appreciating eyes had been
there to see.

But it is not with the poetic side of infancy
that we are here specially concerned.   We
have to look on the opening germ of intelligence
from the colder point of view of science.   Not
that the savant need be insensible to the
æsthetic charm of his subject.   A botanist
*ought* perhaps to feel something of the rich
store of loveliness which lies enclosed within
the tiny confines of a wayside flower.   Scientific
curiosity often leaps into full and vigorous life
under the genial vivifying influence of a glowing admiration.   And a man who has a keen
eye for all the pretty and humorous traits of
infant life is all the better qualified for a close
scientific observation of its processes.   Only

that in this case the æsthetic interest must be
subordinated to the scientific.

The science which is specially concerned
with the baby mind is Psychology. It is only
the psychologist who can pretend to record
and interpret all its strange ways. And on
the other hand, the domain of infant life is of
**peculiar interest to** the psychologist. True, he
can study in other ways the manner in which
the human mind behaves, and the laws which
bind together its sequent movements. **He has**
a mind of his own, which is directly accessible
to his internal vision; and there are the minds
of his friends and acquaintances, about which
he can know a good deal too, always provided
that they are quite open and confiding. Still,
he cannot dispense with the young unformed
minds of infants. His business, like that of
all scientific workers, is to explain the complex
in terms of the simple, to trace back the final
perfectly shaped result to the first rude beginnings. In order to this, he must make a careful
study of the early phases of mental life, and
these manifest themselves directly under his
eye in each new infant.

Some of the gravest questions relating to
man's nature and destiny carry us back to the
observation of infancy. Take, for instance, the
warmly-discussed question, whether conscience
is an innate faculty—each man's possession
anterior to and independently of all the ex-

ternal human influences, authority, discipline,
moral education, which go to shape it ; or
whether, on the contrary, it is a mere outgrowth
from the impressions received in the course
of this training. Nothing seems so likely
to throw light on this burning question as a
painstaking observation of the first years of
life.

This, however, is not the whole of the significance of infancy to the modern psychologist.
We are learning to connect the individual life
with that of the race, and this again with the
collective life of all sentient creatures. The
doctrine of evolution bids us view the unfolding
of a human intelligence to-day as conditioned
and prepared by long ages of human experience,
and still longer cycles of animal experience.
The civilized individual is thus a memento, a
kind of short-hand record of nature's far-re-
ceding work of organizing, or building up living
conscious structures. And according to this
view the successive stages of the mental life of
the individual roughly answer to the periods
of this extensive process of organization—vege-
tal, animal, human, civilized life. This being
so, the first years of the child are of a peculiar
antiquarian interest.

Here we may note the points of contact of
man's proud reason with the lowly intelligence
of the brutes. In the most ordinary child we
may see a new dramatic representation of the

great cosmic action, the laborious emergence of intelligence out of its shell of animal sense and appetite.

Yet it must not be supposed that the interest here is wholly historical or archæological. For in thus detecting in the developmental processes of the child's mind an epitome of human and animal evolution, we learn the better to under-stand those processes. We are able to see in such a simple phenomenon as an infant's respon-sive smile a product of far-reaching activities lying outside the individual existence. In the light of the new doctrine of evolution, the early period of individual development, which is pre-eminently the domain of instinct,—that is to say, of tendencies and impulses which cannot be referred to the action of the preceding circum-stances of the individual,—is seen to be the re-gion which bears the clearest testimony to this preparatory work of the race. It is in infancy that we are least indebted to our individual exertions, mental as well as bodily, and that our debt to our progenitors seems heaviest. In the rapidity with which the infant co-ordinates external impressions and movements, as in learning to follow a light with the eyes, or stretch out the hand to seize an object, and with which feelings of fear, anger, etc., attach themselves to objects and persons, we can plainly trace the play of heredity—that law by which each new individual starts on his life

course enriched by a legacy of ancestral experience.

Viewed in this light, infant psychology is
seen to be closely related to other departments
of the science. To begin with, it has obvious
points of contact with what is known as the
psychology of race (Völkerpsychologie). The
first years of the child answer indeed to the
earliest known stages of human history. How
curiously do the naïve conceptions of nature,
the fanciful animistic ideas of things, and the
rude emotions of awe and terror, which there
is good reason to attribute to our earliest
human ancestors, reflect themselves in the language of the child! It is probable indeed
that inquiries into the beginnings of human
culture, the origin of language, of primitive
ideas and institutions, might derive much more
help than they have yet done from a close
scrutiny of the events of childhood.

Again, it is evident that the psychology of
the infant borders on animal psychology. The
child's love of animals points to a special facility
in understanding their ways; and this, again,
indicates a certain community of nature. The
intelligence of children and of animals has this
in common, that each is simple and direct, un-
encumbered with the fruit of wide comparison
and abstract reflection, keen and incisive within
its own narrow compass. Both the child and
the brute are exposed by their ignorance to

similar risks of danger and deception; both show the same instincts of attachment and trustfulness. And so a study of the one helps the understanding of the other. The man or woman who sees most clearly into the workings of a child's mind will, other things being equal, be the best understander of animal ways, and *vice versa.*

There is one particular aspect of this relation between infant and animal psychology which calls for special notice. The baby contrasts strongly with the young of the lower animals in the meagreness of its equipment for life. Though, as observed above, the child reaps the heritage of the past in instinctive germs of capacity, these are far less conspicuous, far less perfect and self-sufficing than the unlearnt aptitudes of young animals. The young chick seems able to co-ordinate the movements of its head with visual impressions so perfectly from the very first that it can aim with accuracy at so small an object as a grain of corn. The young kitten displays quite an experienced and mature hostility to the hereditary foes of its species. There is nothing corresponding to this in the case of human offspring. The baby has to begin life in the most pitiable state of helplessness. For a year and more he cannot execute one of the most important and wide-spread functions of animal life, namely locomotion. And this prolonged period of helplessness

*b*

has a deeply interesting significance from the point of view of the evolutionist. The backwardness of the human offspring, as compared with the forwardness of the animal, is only a striking illustration of a general law or tendency of evolution. As creatures rise in the scale of organization they have to adapt their actions to a wider and wider variety of circumstances. and actions of the environment. In the lower grades of animal life there is much more sameness and routine just because there is much more simplicity. In the higher grades, actions, having to adapt themselves to more complex and changeful surroundings, are more varied, or undergo more numerous and extensive modifications : contrast the actions performed by the bee in obtaining its food with those carried out by the fox. And the capability of thus varying or modifying actions is the result of individual experience and education. Hence, as the variability of the actions of life increases, so does the area of individual learning or acquisition, as distinct from that of inherited aptitude or instinct. And since the range and variability of human actions are immeasurably greater than those of the most intelligent animal performances, we find that the infant is least equipped for his earthly pilgrimage and has most to do in the way of finding out how to live.

And here we seem to touch on the more

practical side of our subject. To the helplessness of the infant there correspond those instincts of tendance, protection, and guidance
which, though discernible in the lower animals,
are only highly developed in man ; and which,
while they are seen most conspicuously in the
human mother, are shared in by all adults, and
underlie the long and tedious processes of
education. It is not only the theoretic psychologist who needs to study infantine ways ;
it is the practical psychologist, that is to say,
the educator. The first three or four years
of life supply the golden harvest to which every
scientific educationist should go to reap his
facts. For the cardinal principle of modern
educational theory is, that systematic training
should watch the spontaneous movements of
the child's mind and adapt its processes to
these. And it is in the first years of life that
the spontaneous tendencies show themselves
most distinctly. It is in this period, before
the example and direct instruction of others
have had time to do much in modifying and
restraining innate tendency, that we can most
distinctly spy out the characteristics of the
child. It is the infant who tells us most unmistakably how the young intelligence proceeds
in groping its way out of darkness into light.
It is an historical fact, that the supreme necessity in education of setting out with training
the senses and the faculty of observation, was

discovered by a close consideration of the
direction which children's mental activity spontaneously follows.  By sitting at the feet of
nature and conning the ways of untaught
childhood we may learn that all the essential
functions of intelligence,—separation or analysis, comparison, discrimination, etc.,—come into
play under the stimulating force of a strong
external impression.  In the act of holding and
looking at its brightly-coloured toy the infant
is already showing himself to have a distinc-
tively human mind, and to be on the road to
abstract reflection or thought.  It is during
that prolonged gaze that the first rude tentatives
in distinguishing and relating the parts and
qualities of things are effected.  And the object-
lesson, properly conceived, is nothing but a
methodical development of the mental processes
which are involved in every serious effort of
infantine inspection.

Nor is it only on the intellectual side that
this study of the infant mind is of moment to
the teacher.  It is in the first three or four
years of life that we have the key to the emotional and moral nature of the young.  If we
want to know how a child feels about things,
what objects and articles bring him most pleasure, we must watch him at his self-prompted
play and overhear his uncontrolled talk.  It
seems self-evident indeed that if the teacher is
to adapt his method of training so far as may

be to the tastes and predilections of the pupil, he must have made a preliminary study of these in their unprompted and unfettered expression. If the study be deferred to school life it will never be full or exact. The artificial character of even the brightest school surroundings offers too serious an obstacle to the free play of childish likings.

Enough has been **said**, perhaps, to show that the observation and interpretation of the infant mind are at once a matter of great theoretic and practical importance. And now comes the question : By whom can this line of research be best pursued ? The conditions of success are plainly two : (1) proper qualifications for the work, and (2) ample opportunity.

1. With respect to the first condition, it has already been suggested that a good observer of childish ways must combine a number of intellectual and moral excellences. He must, to begin with, be a painstaking and exact observer. He must be determined to see children as they actually are, and not to construct them out of his own presuppositions. And this implies a mind trained in observation, and a certain scientific rigour of intellect. Yet this is clearly not enough, for many an excellent observer of other domains of nature might prove a very sorry depicter of infant traits. The close habitual concentration of the mind on things so trivial, to robust common sense,

as baby whims and oddities, presupposes a selective emotion, a strong loving interest in this particular domain of natural fact. And this, again, implies that the observer should be touched by that enthusiasm for childhood which shows itself as a kind of consuming passion in men like Pestalozzi and Froebel.

Nor is this all. The infant mind cannot be seen, but only divined. Every movement of the tiny hands, every modulation of the baby voice, is as meaningless as sounds of an unknown tongue, until the interpretative work of imagination is added. And it is just in the ability thus to construe the external signs of infantine feeling and thought that so many otherwise good observers fail. Nothing, perhaps, has been more misunderstood than childhood. Few have the retentive memory of their own early experiences which would at once put them *en rapport* with the mind they are observing. And few have the disposition to seriously endeavour to think themselves into the situation and circumstances of the child, casting aside their own adult habits of mind, and trying to become themselves for the moment as little children.

Neither the close observation nor the careful interpretation of children's words and actions can be counted on where there is not love and the habitual companionship which grows out of love. The man to whom children will reveal themselves, is not he who is wont to look on them as a nuisance or a bore, but he who finds them an amusement and a delight, who likes nothing better than to cast aside now and again the heavy armour of serious business, and indulge in a good childish romp. Understanding of the child's mental workings, his own peculiar *manière de voir*, his standard of the importance of things, and so forth, presupposes a habit of steeping the mind in the atmosphere of child-life.

2. It follows that a complete qualification for the office includes the second condition, namely ample opportunity. Nobody ever acquired the art of reading the book of child-nature who had not enjoyed full opportunity of observation. When, however, we consider the first year or two of life, we see that opportunity is necessarily greatly restricted. Beyond the mother, nurse, and perhaps the doctor, who is there that is privileged to watch the first tremulous movements of the baby-mind ?

And here the thought naturally occurs, that the mother is the person specially marked out by nature for this honourable task. She is, or ought to be, the one who comes into closest contact with the baby, and gives it the first sweet taste of human fellowship. She too has, or ought to have, the liveliest interest in the child, the absorbing interest of idolatrous maternal love. She, we all cheerfully grant,

will grudge no effort spent in divining the
direction of those first obscure baby impulses,
the form of that first unfamiliar baby thought.
But has she the **other** qualifications—the mind
severe in its **insistance** on plain ungarnished
fact, trained in minute and accurate observation
and in sober methodical interpretation ?  Here
our doubts begin to arise.  Few mothers, one
suspects, could be trusted to report in a perfectly
cold-blooded scientific way on the facts of infant
consciousness.  The feelings which rightly tend
to baby-worship would, one feels sure, too
often lead to an arbitrary limitation of the area
of fact, to confusion of what is actually observed
with what is only conjecturally inferred, to
exaggeration and misrepresentation.  The very
excellences of maternity seem in a measure to
be an obstacle to a rigorous scientific scrutiny
of babyhood.

The doctor would of course be much more
likely to possess the scientific qualifications for
this office of baby-interpreter.  And medical
men have been known to throw themselves
into the work.  At the same time it is obvious
that a doctor's preoccupation of mind with the
physical state of the infant would necessarily
interfere with a close attention to psychical
traits.  And at best he could only obtain, by
his direct observation alone, a few fragmentary
results.

And here, perhaps, we may do well to think

of another possible candidate for our post. The father can, it is evident, find an ampler opportunity than the doctor for a continuous systematic observation of his child. No doubt he will have obstacles put in his way. It is not improbable that the nurse may assert her authority and set her face resolutely against a too free intrusion of man's footsteps into the woman's domain. Still these obstacles may by judicious cajoling be greatly reduced in size, if not altogether removed. In most cases, it may be presumed, he will have a moderate paternal sort of interest in the doings of his tiny progeny. And his masculine intelligence will be less exposed to the risk of taking a too sentimental and eulogistic view of the baby mind.

The father cannot, however, hope to accomplish the task alone. His restricted leisure compels him to call in the mother as collaborateur. Indeed, one may safely say, that the mother's enthusiasm and patient brooding watchfulness are needed quite as much as the father's keen analytic vision. The mother should note under the guidance of the father, he taking due care to test and verify. In this way we may look for something like a complete record of infant life.

It is satisfactory to find that fathers are waking up to a sense of their duty in this matter, and are already laying the foundations

of what may some day grow into a big biographical dictionary of infant worthies.   The
initiative, as might have been expected, has been
taken by men of scientific habits and tastes.
Physicians, naturalists, and psychologists have
co-operated in this useful parental work.  Among
physicians may be named Tiedemann, Sigis-
mund, and Löbisch.   **Among** naturalists figure
**the** names of Darwin and Professor Preyer.
And the psychologists are represented by M.
Taine, M. Perez, Mr. F. Pollock, and others.[1]

The volume which is here presented in an
English garb, is from the pen of one who
**combines** considerable physiological and psy-
**chological** knowledge with a practical interest
in education.   M. Perez is best known perhaps
as a writer of **pædagogic literature.**  He has
written a volume, as **well** as occasional articles
on distinctly pædagogic themes, and in addition
**to this** has edited writings of other pædago-
gists.

The peculiarity of this record of the first
three years of the child, is that it is not a bio-

---

[1] Reference to the bibliography of the subject will be
found in Preyer's *Die Seele des Kindes,* cap. 19.  The observations of Darwin, Taine, Pollock, and others are recorded
in *Mind,* vol. ii. pp. 252, 285 ; vol. iii. p. 392 ; vol. vi.
p. 104.   I may also refer to two articles of my own, one on
*Babies and Science,* in *The Cornhill Magazine,* May, 1881,
and one on *Baby Linguistics,* in *The English Illustrated
Magazine,* November, 1884.

graphical sketch.   M. Perez, so far as we can
judge, has made special note of the progress
of one or two favourites, but his record is a
wide and comparative one.   This gives it its
peculiar utility.   Each mode of chronicling the
events of child-life is valuable—the careful chro-
nological report of a single child's development,
as that of Tiedemann, Darwin, Preyer, and
others, and the larger survey of facts which
comes from the observation of a number of
children and the averaging of the results
reached, as in the work of M. Perez.   It may
be added that our Author appears to have enjoyed very exceptional advantages in finding
out the ways of infants.

   The obvious defect of a single biographical
record is, that it cannot be taken as typical.
As every mother of a family knows, children
manifest striking differences from the very
beginning of life.   Indeed, it is not too much
to say, that the child shows its individuality
the very first day of its post-natal existence, in
the way it takes to the nutriment provided by
nature.   The differences of mental precocity in
infants are very striking too.   It is one merit
of the present volume, that it presents us with
a wide variety of childish character.   In some
places we have a distinctly precocious trait
recorded, as for example in the odd display of
quasi-pity by a child of sixteen months, at
the sight of an adult undergoing a douche bath

(p. 80).   *En revanche*, we have in other places instances of quite commonplace achievement, if not of decided backwardness, as when it is recorded that a child of eleven months was able to understand a number of words and "even a few little phrases" (p. 238).   It is only by taking the dull and the clever infants together that we are able to reach the idea of an average typical development.

M. Perez combines, I think, in a very happy and unusual way, the different qualifications of a good observer of children.   He has the first condition—loving interest, and the clear sympathetic insight which grows out of this.   Even the much-neglected dreams of children are a matter of concern to him, and receive illumination from his bright intelligence.   Nor is he without a quick sense of the poetic charm of babyhood.   Some of the stories he tells us are as fresh and delightful as idylls.   They transport us into the very atmosphere of unconventional child-nature.   At the same time he never allows his sentiment to get the better of him.   He is before all other things *savant*, and as such he exposes the unlovely side of infancy in a most merciless fashion.   The account of the little ones' fierce angers, petty jealousies, and brutal insensibilities to the sufferings of others will perhaps horrify some readers, who are accustomed to think of them as having only a divine or angelic side; but they will be appreciated

by every one who cares more for the **accuracy**
of facts than for their conformity to our wishes
**and** fancies.

Another prominent feature of this work, is
its clear recognition and appreciation of the
bearings of evolution on the facts of child-life.
M. Perez is evidently an **ardent** evolutionist,
**and makes** excellent use of the new doctrine in
explaining what he sees. This feature gives
the air of newness **to** the volume. The
reader feels that he **is** listening to one who
is fully abreast **of** the latest developments of
science.

With this feature may be coupled another,
namely, the ample **reference to animal** psychology. M. Perez illustrates **the observation
made** above, that interest in children has a
**close** kinship with interest in animals. He has
himself **been** a careful observer of domestic
animals, **and** his references to his kittens are as
delightful to the imagination as they are helpful
to the understanding.

As remarked, **M. Perez looks at** the **infant**
from an educator's point **of view.** He knows
very well that education begins from the cradle,
and his book abounds with practical hints on
the proper way of training the very young.
His kindly nature is quick **in** detecting the
woes of childhood, and eloquent in pleading
for their mitigation. Instance what is said
about the wickedness of deceiving children

(p. 98).   At the same time, the pædagogic
intention is never obtruded unpleasantly on
the reader's notice.   It is by way of passing
suggestion, rather than of elaborate enforce-
ment, that he aims at making this study of
facts a practical guide to the mother and the
teacher.

A last feature of this volume which is deserv-
ing of mention, is its thoroughly French form
and style.   The reader feels at every page that
he is listening to a Frenchman who knows how
to shape his materials, give order and arrangement to his exposition, light it up with per-
tinent illustration, and adorn it with the graces
of style.   While in places the Author ventures
a few steps into the darker recesses of metaphysical psychology, he never long forgets that
he is writing a popular work.   And he has
succeeded in producing a volume which, while
it will be of special interest to the scientific
student, will attract the general reader as well.

It may not be superfluous to say perhaps,
what I feel sure the Author himself would
endorse, that this volume makes no pretension
to be a final and exhaustive study of its subject.
A complete theory of the infant mind will need
to be built up by the combined efforts of many
observers and thinkers.   In the region of psychology, much more than in that of the physical
sciences, repetition of observation and experiment is needed to check and verify the results

of individual research.    The secrets of infancy
will only be read after many pairs of eyes have
pored over the page.    Though, as observed,
M. Perez has made his studies unusually wide,
it may be reasonably doubted whether in some
cases he does not give exceptional instances
as typical and representative.    Certain it is
that his notes respecting the first appearance of
sensations, *e.g.* those of taste and smell, of the
perceptions of distance, etc., of the movements
of grasping objects, and so forth, differ in some
important respects from those of other observers.    In certain particulars, too, this volume
is less full than some other records, notably
that of Professor Preyer's *Die Seele des Kindes*,
which, as it was published after the work before
us, is not referred to.    Hence the student who
wants to be quite abreast of the present results
of research will do well to read other records
in company with this.    This circumstance however does not in the least detract from the
value of *The First Three Years*, as a rich mine
of facts, and one of the fullest, if not indeed
the very fullest, monograph on its subject.

In conclusion, I would express the hope
that this pleasant volume will stimulate many
an English parent to new individual research
in the same promising field.    After reading it
carefully, any person of ordinary intelligence
will have learnt something about the things to
be looked for, and the way in which they are

to be looked for.   And I trust I am not going
*ultra vires* in reminding my readers that there
is an English journal of psychology, the Editor
of which has proved his readiness to publish
contributions to the young and promising science
of baby-lore.

JAMES SULLY.

# CHAPTER I.

## THE FACULTIES OF THE INFANT BEFORE BIRTH.

CAN the fœtus be regarded as belonging to psychology? Psychologists have no doubt whatever on the subject, for they will not admit that any organism exists without functions in a greater or less stage of development, and they see in the fœtus a rough sketch, as it were, of the entire system of the organs of sensibility. No doubt, either, is expressed on the question by philosophers of the experimental school, who accord as much importance to the unconscious as to the conscious life of the mind, and who regard the apparently automatic movements of the intra-uterine stage of existence as evident manifestations of sensibility of some sort. We will begin by quoting the opinion expressed by Dr. Luys concerning initial sensibility.

"In the first phases of fœtal life it is very difficult to fix definitely at what epoch sensibility manifests itself as a motor force; nevertheless, from the fourth month we can observe that the nervous system begins to react and reveal the vitality of the different apparatuses of which it is made up. We know, indeed, that from this period the fœtus is sensitive to the action of cold, and that we can develop its spontaneous movements by applying a cold hand to the abdomen of the mother. We know also that it executes spontaneous movements to withdraw from pressure that constrains it and brings its sensibility into play. We may then legitimately conclude that here we have the first gleams of awakening sensibility, which from this period is

B

transmitted through its natural channels by the nervous system and already regulated in the manner in which it will subsequently manifest itself throughout the organism." [1]

If we admit with this writer the persistence in the nerve cellules of the vibrations which first set them in motion, and, as regards intellectual activity, the persistence in these vibrations of memories and ideas, unconscious though they be, we shall understand the great interest which psychologists must feel in the study of the functional action, whatever its nature, of psychological life before birth.

M. Ribot, a philosopher of the experimental school, admits the possibility of researches useful to psychology being made on the subject we are treating of, but only from the point of view of our knowledge of the unconscious life. The first forms of unconscious life must be sought for in the fœtal life—a subject full of obscurity, and very little studied from the psychological point of view. We may hold, with Bichat and Cabanis, that though the external senses are in the fœtus in a state of torpor, and though, in the constant temperature of the amniotic fluid, the general sensibility of the fœtus is almost null, still its brain has already exercised perception and will, as seems to be evidenced by the movements of the fœtus during the last months of pregnancy. [2]

Professor Kussmaul goes still further. He admits " that the child can, even before birth, experience certain sensations and acquire certain aptitudes by means of the sense of touch aroused in it through contact with the matrix which surrounds it, as well as by the sensations of hunger and thirst excited by the amniotic fluid which it swallows. Thus, at this period already, the intelligence of the child would begin to be developed, although very imperfectly."

What, we ask, do these modifications of the mental life before birth amount to? May we hope that experimental analogy will one day enlighten us on this delicate and com-

---

[1] Luys. *Le Cerveau et ses Fonctions.* See English translation (International Scientific Series), p. 126.
[2] Ribot, *L'Hérédité.* See English translation, p. 227.

plex question ? Will not the comparative anatomy of the fœtal brain, the child's brain, and the adult brain some day inform us whether the primitive sensations are or are not arrested before they reach the nerve fibres which help them to transform themselves into perceptions ? The organs of sense, although as yet incompletely finished, do they not nevertheless produce local peripheric reactions, sensitive vibrations, or veritable sensations, which are a kind of rudimentary exercise and training to the parts acted on.

What is the action of external or internal life on these semi-formed, semi-active, perhaps vaguely conscious souls. If the movements of the fœtus were all mechanical and unconscious, would its sensations end only in dull excitations of the sensibility—the importance of which it is as unreasonable to deny as to exaggerate ? Since the deviation of the tenth part of a millimeter, sustained by a nerve fibre during pregnancy, exercises a considerable influence over the contexture of any one organ, the impressions received during fœtal life—whether regular or accidental—and although working on organs which are still imperfect and unexercised, must nevertheless have an influence on the general constitution of the mind, as well as on its special dispositions, which it will no doubt be possible to determine within certain limits. These are the first rays of spiritual life whose dawn is the moment of birth. It is the first page of a book which the psychologist should neither despair of being able, nor be in too great haste to decipher.

What is the nature or the intensity of the sensations which we suppose the fœtus to experience ? They are probably sensations already localized and differentiated, of visceral comfort or discomfort, muscular, cutano-thermal, tactile, or sapid sensations ; these latter less distinct because they are less varied than the vital and muscular sensations ; we may even suppose that, in spite of the imperfect development of the auditory apparatus, the vibrations of sound, transmitted through the abdominal partition or chamber, would produce on the parts already completed concussions corresponding to a rudimentary and general sensation of hearing. From what we have been able to gather, however, as to the nature of these hypothetical sensations, we may infer that they are not very distinct or intense.

Contemporary physiology recognises generally in sensation only the result of the modification of the organs which takes place in correspondence to each modification of the surrounding medium. The intensity of sensation is in direct ratio to the resistance offered by the fibres of sensation to the action of external impressions. It is the repercussion of this action on the sensorium which produces sensation. The nerve of an adult is able to bear greater excitation than that of a young child, and requires for the production of a given sensation a minimum excitation stronger than the minimum excitation which would produce the same sensation in the case of the child. But the extreme adaptivity of the infant organization to its surroundings, and the small resistance which it offers to external excitation, renders the sensational susceptibility considerably less than at a more advanced age. Extremely open to new impressions, children resist them but little, easily get accustomed to them, and also, for that very reason, do not feel them long. And all this applies still more strongly to the fœtus. The relative activity of the visceral functions, the extreme tactile sensitiveness of the skin,—partly the result of its yet imperfect state,—the already considerable development of muscular sensibility, may all cause the fœtus to experience distinct and acute sensations, under the influence of pressure, of heat or cold, or of internal modifications of the organization.

What do we know, however, concerning the degree of actual vitality inherent in the organs and the nervous centres at the period of which we are treating. In a word, does the fœtus really feel and receive the sensations which we suppose it does?

Most physiologists are, like Wirchow, of opinion that the new-born child is simply "a vertebrate animal," that its unconscious sensations are nothing more than automatic reflex actions, that its sensations and movements have no echo in the centres of sensational and motive ideality. Does this mean that these two encephalic centres are inactive in the fœtus? Our scientific knowledge of the fœtal brain, as far as I know, can teach us nothing decisive on this point, and the brain of the new-born child is still fallow ground, if we may judge from the small amount of information to be found on this subject in the most recent French works on the brain in general. We quote from Dr. Charlton Bastian some interesting passages on the development of the brain during fœtal life.

" By the end of the fifth month the growth of the cerebral hemispheres has been so great that they completely cover not only the corpora quadrigemina but also the now larger cerebellum. . . ."

" In the remaining important section of intra-uterine life, from the sixth to the end of the ninth month, the developmental changes in the cerebrum are much more marked than they are in the cerebellum. The walls of the cerebral hemispheres become thicker, and there is a proportionate diminution in the capacity of the 'lateral ventricles,' the three 'horns' of which now become quite distinct. The corpus callosum assumes a more horizontal direction, whilst it increases both in thickness and in length. . . ."

" During the sixth month a surprising development of the fissures and convolutions takes place, so that early in the seventh month all the principal of them are distinctly traceable. . . ."

" During the attainment of this degree of convolutional complexity, some important changes have been taking place in the relative development of the different 'lobes' of the brain. At the seventh month the parietal lobe (which at birth showed proportionately fuller development) is notably small, while the frontal or temporal lobes are large. . . ."

" According to S. Van der Kolk and Vrolik it appears that in their relative proportions, the lobes of the brain in a new-born child hold just the mean between those of a chimpanzee and of an adult man. In the adult orang, however, the same proportion obtains between its different lobes and those of the new-born child. . . ."

" In regard to the microscopical characters of the fœtal brain, one brief but important statement deserves to be recorded. According to Lockhart Clarke : 'In the early fœtal brain of mammalia and man the structure [of the cerebral convolutions] consists of one uninterrupted nucleated network. As development advances, separate layers may be distinguished.'[1] But even in these layers there are only to be recognised, ' roundish nuclei connected by a network of fibres,' or, in other parts, groups of more elongated nuclei, in place of the distinct but differently shaped nerve cells with inter-connecting processes which are the prevailing and characteristic constituents of the cerebral convolutions in their developed condition."[2]

As far as these superficial indications allow of any conclusion, we see that if the development of the superior hemispheres appears to be ripe enough for perception, emotion, and volition, the defective consistence of their constituent elements indicates, at least during the uterine period, a very limited functional power of the psychical faculties. But it does not follow from the fact that the development of the cerebellum, the organ of co-ordinate movements, is not so advanced as that of the cerebrum, that the centres of ideality and of motive volition have not already entered on their functions before birth.

Psychologists ought then to be very careful of pronouncing judgment in a matter where physiologists of the highest authority are still so much in the dark. I may add, that the latter not being able to subject new-born babies to the murderous experiments of vivisection under anæsthetics, we have at the present moment but little to expect from their direct sources of information on the cerebral functions of young children—and à fortiori of the fœtus. At the most, as we shall show further on, they may contribute to infant psychology some indirect information from their experiments on new-born animals. Every age has its comparative psychology.

For the present, however, we may fairly assume from

---

[1] *Notes of Researches on the Intimate Structure of the Brain.* Proceedings of Royal Society, 1863, p. 721.
[2] Charlton Bastian, *The Brain as an Organ of Mind* (Internàtional Scientific Series), ch. xix., pp. 341–346.

analogy, that a long time before it is born a child will have become acquainted both with pain and pleasure, in so far, *i.e.*, as its gradually developing organs have allowed the passage of the impressions which normally produce these sensations. It will also have experienced a great number of lesser sensations, which, though they may have been almost indifferent to it, will nevertheless have had some sort of echo in its already formed consciousness. Most of these sensations, from being produced without the concurrence of the higher brain action, will be confused and indistinct. Others, however, will be clearer, and result in true rudimentary perceptions—vague perceptions, *i.e.*, and not localized, and without any other connecting link than that obscure and innate sentiment of personality which is buried in I know not what mysterious corner of the encephalus. These perceptions are, as it were, interior and subjective. We shall have occasion to return to the subject of perception during foetal life in the chapter devoted to the perception of the new-born child.

## II.

### THE FIRST IMPRESSIONS OF THE NEW-BORN CHILD.

The child has already made acquaintance with pain before coming into the world, and we cannot tell how much more the hard labour of birth may have caused it to suffer. There is no doubt, however, that the moment when it first enters into relation with external realities is a very painful one. As soon as its head comes in contact with the air, this fluid pours in torrents down the delicate tissues of the respiratory organs; and the progressive succession of movements of inspiration and expiration, which are the beginning of pulmonary respiration, are not effected without painful shocks. This is why a new-born infant utters sounds similar to those produced by suffocation. It is only slowly and after several days' time, that the little being gets accustomed to the atmosphere which surrounds it. When it first comes into the world, unfurnished with

the power of resistance which it acquires later, its delicate
skin is suddenly enveloped in an atmosphere which is icy
cold compared to that which it has just left. Cold is the
most serious enemy to new-born children. We know with
what care animals hasten to warm their little ones with the
heat of their own bodies. The first operation of washing
is another source of suffering, for, whatever care be taken
and however rapidly it be performed, the towel and the
sponge, even though of the finest texture, must irritate the
tender skin of a new-born child. The freedom of movement, too, which its limbs have gained is in itself a source
of pain, after the soft pressure they have been accustomed
to ; and the most gentle handling of the nurse or mother
are torture to its delicate frame, still suffering from the
struggle of coming into life.

All its senses are battered by repeated shocks of strange
impressions, and its wailing cries indicate how painfully
these are felt. A new-born infant is both deaf and blind,
but the rays of light strike none the less on its eyes with a
sense of shock, and the waves of sound dash all the same
against its tympanum.

Can we then wonder that the child's entrance into life
should be accompanied by those plaintive cries so well
described by the Latin poet. "A child at its birth, like
a mariner cast ashore by the angry waves, lies prostrate on
the earth, naked, speechless, destitute of all the aids to
existence, from the moment when it reaches the shores of
light, torn from its mother's bosom by the efforts of nature ;
and it fills the place it has entered with dismal wailings.
And such distress is but natural ! There lies before him
of traverse a life afflicted with bitter woes."[1]

The child, however, has no presentiments of the sorrow or
the joy which will make up his longer or shorter career; but
the individual life that has become his is evidently a cause
to suffering to him. Like new-born animals, he is restless
and wailing under the influence of all the new impressions

---

[1] Lucretius, *De Natura Rerum*, lib. v., ver. 223, etc.

which irritate his keen susceptibilities; and it is strictly true
that children make acquaintance with external life through
suffering. Their entrance into the world is as painful as
their exit; and this is the reason why so many children,
exhausted by all their efforts and sufferings, fall after birth
into a kind of corpse-like rigidity. A new-born infant
shows no immediate desire for food. This is probably
owing to exhaustion and the strong need of sleep, and
perhaps also to the powerful action of the air, which at
once begins to transform the venous blood into arterial
blood.

"It is not till after several hours that the need of nourishment first makes itself felt. The sensations of hunger and
thirst, both equally new, recall the child from the oblivion
of sleep, and it awakens with a cry. Maternal tenderness
responds instinctively to this appeal, and offers its first gift.
The child now learns the delight of moistening its lips with
a sweet and pleasant liquid, imbibed from the breast on
which its head reposes so softly; and the first sufferings
experienced at its *début* are quickly effaced by this first
sweet pleasure. The satisfied child falls asleep again on
its mother's bosom, with the feeling of comfort which
satiety produces, and seems once more to return to the
isolated life which was its normal condition in the mother's
womb, and which its organism has not yet lost the habit of.
It reawakens from this slumber every time that the need of
nourishment returns to trouble its repose."[1]

Thus, though the child thrown naked on the earth soon
finds there the soft warm pillow of its mother's breast;
though nature herself manages for it the transitions which
lead it gently up to the complete possession of sight,
hearing, and locomotion, the struggle for existence begins
nevertheless at birth. Hunger, thirst, cold, difficulty of
respiration and of digestion, tactile and muscular sufferings,
irritating impressions on the organs of sight and hearing,
all these await its awakening from each of its deep

---

[1] Richard de Nancy, *Education Physique des Enfants.*

slumbers.   And however little conscious and vivid these sensations may be, they none the less affect the delicate and still imperfect organs.   This accounts for the great difficulty there sometimes is in rearing little children, and the frequency of nervous or pectoral affections which carry off so great a number.   Moreover, if it is reasonable to judge of the welfare of a child by the strength of, and the manner in which, its various functions operate, we may fairly attempt to determine the condition of a new-born infant by this one action of feeding.   During the first few days babies suck feebly and without energy, and quickly weary of the process.   At this period they are suffering from various operations of nature, and up to the third day there is a diminution of between three and four ounces in their weight ; and it is not till the sixth or seventh day that they get back to what they were at birth.   Thus we conclude that a child's general well-being corresponds to the degree of development of its organs and their power of adaptation to their new surroundings and conditions.

## CHAPTER II

## MOTOR ACTIVITY AT THE BEGINNING OF LIFE.

As the first manifestations of psychic life in a new-born
child are shown in movements which are more or less
automatic, it will be well for us first to study the nature of
these movements, which in adults are the expression of
psychological phenomena. From the very first days we
perceive in infants a sort of *motor activity*, general and
indefinite, which is due to external or internal excitations,
themselves also for the most part vague and undefined.
We recognise in all this activity a spontaneous tendency of
the nervous centre to spend its superabundant energies in
muscular force ; but we can also see in some of the actions
the expression of pre-established association between
certain movements and certain sensations, agreeable or
otherwise. For instance, the vague incoherent movements
of the arms, legs, and facial muscles, which young babies
make, as if trying to escape from the pressure of their
clothes, or struggling against some painful state of their
system ; the aimless movements which their arms perform,
striking right and left without any definite object ; all these
belong to the first class of indefinite reflex actions. From
the first also, the mouth will seize and suck eagerly a finger,
or anything else held before it ; the fingers close mechanically round any object with which the palm is touched ;
" Like the leaves and flowers of a sensitive plant whenever
they are touched by a foreign substance." [1] The eyes blink
uneasily, as if endeavouring to accustom themselves to the

---

[1] *Mémoire de Tiedemann.*

11

blinding light; if the soles of the feet are touched gently, or tickled with a feather, instinctive movements are at once made to draw away the feet. All these movements, in themselves automatic, but of definite use, belong to the class of general distinct movements. According to Darwin, one of the first reflex actions to be noticed in babies, and one which appears at first sight to have nothing to do with instinct, is the sneezing which accompanies the first act of respiration. The same applies to the shrill staccato screams which are the preliminaries to the action of crying, the pouting of the lips, the wrinkling of the forehead and eyebrows, the contraction of the mouth, by which the lower lip is raised, with a convulsive depression of the corners of the mouth, and finally the sobs and tears, when the child is old enough to produce them. Darwin tells us that in one of his own children he noticed a distinct sob at the age of 138 days. With regard to the effusion of tears, his observations show the period of their first appearance to be very variable. He has never seen them before the twentieth day; once at the sixty-second day; two other times at the eighty-fourth, and once not before the hundredth day. Laughter, which begins so early, is manifestly a reflex action. This universal sign of joy consists in the retraction of the corners of the mouth, with a slight separation of the lips; and we may remark this slight expansion of the anterior muscles of the face, this smiling countenance, even in the faces of dogs and cats when at play. Tiedemann thought that he noticed in his boy five days after his birth "the appearance of laughter without any particular motive, thus most probably without intention or sentiment of pleasure, and simply the result of chance action of the mechanism." Tiedemann's son, however, was altogether remarkably precocious. I would rather quote the authority of Darwin in the case of points which have been scientifically observed by him.

"Those who have the care of young children," he says, "know very well that it is difficult to determine with certainty whether particular movements of the mouth express anything, *i.e.*, whether a child is really smiling." He has scarcely ever observed anything like a real smile before the age of forty-five days ; and the broken, jerky sounds of genuine laughter he has never heard before the sixty-fifth day.

" In this gradual development of the laughing powers we find, up to a certain point, an analogy with what takes place in the case of tears. It seems that in both cases a certain amount of practice is wanted, just as for the production of the ordinary movements of the body, or in acquiring the power of walking. The faculty of screaming, on the contrary, the use of which is very evident, is completely developed at the outset." I have seen smiles on many infants' faces before they were a month old. But this henceforth reflex symptom of joy must not be confounded with certain spasmodic and jerky twitchings of the lips which acute pain produces in infants as well as in grown-up people. All the children of two months old that I have observed have laughed genuinely if they were tickled, or if anything pleased them; and their laughter has been more or less easy and distinct according to their organization. But they were not themselves aware that their smiles or laughter expressed anything, and I have seen very few infants at this age respond with a smile to their mother's smile. An appeal of a more physical nature is needed at this period, an appeal of the voice, or a sensation of physical pleasure. Or if ever they smiled consciously without any apparent stimulus of this kind, the intention was probably a very feeble one.

People who have to do with young babies know how, even before the age of two months, they delight in moving their arms and legs about in their cradles. While waiting for the power of locomotion to develop, the tendency to motion vents itself in incessant exercise of the limbs. And these movements, for the most part vague and indefinite, are accompanied by "sounds, simple and inarticulate it is true, but full of variety" (Tiedemann). Nothing shows more clearly the intimate relation between the motive centres of the limbs and the centres of articulation, than this necessity, both for children and animals, of associating sounds with movements. M. Taine gives an instance of this in a little girl on whom he made observations : " To-

wards the age of three-and-a-half months she used to be
laid out on a rug in the garden, and there, lying either on
her back or her stomach, she would fling her four limbs
about for hours together, uttering at the same time all sorts of
varying cries and exclamations ; nothing but vowel sounds
however, no consonants ; this went on for several months."
But sounds, either articulate or inarticulate, are of no great
importance as far as psychology is concerned, since the
acquired language of gestures and words, as soon as children
understand their expressive value, render these early utterances useless.  It would be very interesting, however, to
study them in the case of children placed, if such a thing
were possible, in such a situation that they would have to
re-invent their language, without other help than the gestures
of their teachers.  The experiment may possibly be made
some day.

But to return to the study of movements, so interesting to
the psychologist, because they are most often the expression
of mental phenomena, which in their turn they again produce.[1]  Movements that are very frequent with children
hardly a month old, are those which they make when they
are held up, with both arms or with only one, with a slight
impulsion up and down, without apparent purpose and in a
more or less marked manner, accompanying the movement
with little clutchings of the fingers.  In like manner, when
their legs are free, they kick them up and down with an
automatic regularity which reminds one of certain convulsive movements of adults.  Must we not recognise in these
movements a discharge of the surplus nervous activity,
which neither has nor seeks any special end, but which
nevertheless has a useful effect on the whole system of vital
functions ?  It seems to me that we may also reasonably
regard them as the effect of latent tendencies eager to
manifest themselves by specialized movements ; and moreover the result is facilitated by these movements, though
they may have no definite aim.  In fact, each nerve and
each muscle produces its own action, more or less regular,

[1] See M. Ribot, *Revue Philosophique*, Oct. 1st, 1879.

when its fibres and centres have reached a sufficient degree of strength or organization ; and further, exercise, which is favourable to the general development of the body, prepares also the way for powers latent in the organization, and which sometimes burst quite suddenly, or at least without apparent gradation, into actuality.

During the three first months the movements of the head from right to left, or left to right, or up and down, which continue for some time to be so uncertain and feeble, assume by degrees a more definite character. In the case of a child of a month and a half only, I have noticed a movement of the head which followed a look cast in my direction, while the ear was bent forward as if listening to me. I have seen a child at the age of seven weeks repeat twice over a face of disgust at the sight of a dose of medicine, accompanying the grimace with movements which later on will signify *No*, but which at the time were only unconsciously repellent. This, however, was a step in advance : an infant, of eight days or a month simply makes a face, without any accompanying movement, to reject anything that is disagreeable to its taste. As for the movements of the eyes which Tiedemann has observed, and which often occur the second day after birth, they can only be considered as purely reflex actions, even at the age of three weeks or a month, since, as I have already said, children are blind during a period the length of which it is impossible to determine. It may be, however, that the vibrations of light excite their eyes to automatic movements which seem, by chance, to be following some object. These movements are perhaps a necessary stepping-stone to sight, for everything tends to the belief in a vague correspondence between light and the tissues intended for vision, since a like correspondence is found where there is little or no visual organ.

"The rudimentary eye, consisting, as in a Planaria, of some pigment grains, may be considered as simply a part of the surface more irritable by light than the rest." [1]

*A fortiori* may this be the case with young babies, who

---

[1] Herbert Spencer, *Principles of Psychology*, p. 314.

have a special organ of sensibility to light.   The automatic movements of the eyes, which are produced by mere irritation, without any sensation of light, contribute nevertheless to the nutrition of the muscular and nervous tissues, the development of which is indispensable to the production of visual sensibility.   As the development of the muscles, the nerve centres, and the motor centres proceeds, as the sensations become distinct and the judgments more extended, the motor centres acquire more and more specialized powers of adjustment.

A child of two months who can distinguish several objects outside himself, and is beginning to have a vague idea of distances, not being able to stretch out his hands and seize distant objects, as he does those near to him, bends his whole body towards them.   At this same age children begin also to have a clearer idea of extent, or rather of localization relatively to the different parts of their bodies; they now only scratch themselves at intervals.   Before the end of the third month, they begin to lift their hands to their faces oftener than before, and a little later the first pains of teething cause their fingers to be incessantly carried to the mouth.   They now use their hands more and to better purpose; the flexion of the hand from the wrist is accomplished perfectly; the fingers have acquired more regular and more varied movements, and we notice efforts at stretching out the arms; this, however, rarely as yet.   A child at this age will also attempt movements of the legs and the thorax to balance himself in an upright position when held up on his feet, and will struggle with arms and knees to climb up to his nurse's face, when she helps him forward on his feet.

In a word, he has gained greater consciousness and mastery of his activity; it affords him more pleasure, and this very pleasure excites him to the use of it.

## II.

### MOTOR ACTIVITY AT SIX MONTHS.

From the fourth to the sixth month, progress is made in numerous ways, but very slowly in comparison to some of the superior animals.   The reason of this is, that the more the will is intended to operate in the control and co-ordination of the movements of the body, the more various and complex will be the special locomotive attainments and the longer the process of training.   A cat at a month old, or a dog at four months, will better use its paws for standing, walking, seizing, or playing, than a year-old child will use its hands and legs for the same purposes.   A kitten, immediately after birth, can drag itself along on its stomach, if not walk; and its efforts and progress, especially in locotion, may be with interest compared to the corresponding progress which takes place so slowly in a child.[1]

It takes a child two years to accomplish stages of progress which a kitten gets over in less than a month.   A child of six months, if lowered rapidly in its nurse's arms, will scarcely put out his arms to save himself from falling.   His triumph is in the sitting posture.   Seated on the floor, surrounded with playthings, of which he shows himself a jealous master, his hands, arms, and fingers accomplish many delicate and varied movements of which a cat or a dog of a year old would not be capable.   His activity, doubled now by curiosity, and stimulated to the highest pitch by emotional sentiments of all sorts, makes him happier and happier, and seems to him so great a necessity, that a quarter of an hour of relative inactivity weighs on him as much as a whole day of *ennui* on a grown-up person.

At this period a child gets an immense amount of pleasurable sensation of all kinds, muscular, intellectual, and moral, from its first attempts at walking and talking and imitating all the different gestures of the people around ; and all these progressive steps are so much the more keenly relished by its sensitive personality that they are not the result of so many conscious efforts or of gradual and intentional evolution, but more often come quite suddenly into operation, according as the development of the different organs which produce them proceeds.   The sudden apparition of these

---

[1] See *Mes Deux Chats, Fragment de Psychologie Comparée*, broch. in-12, chez Gernier Baillère, 1881.

C

forces, of which possibly a vague presentiment may have
been felt throughout the organism, but which have not been
intentionally called up, affords a child the constant delight
of fresh surprises. They are treasures which he wishes to
share with all around ; he will repeat, untiringly, some new
movement of which he has just discovered in himself the
power—a special movement of the hands or legs, for instance ; and he will endeavour to apply it to any and every
purpose, just as the first significant articulations which he
has learnt to utter serve him for general terms to designate
objects which perhaps have only a very distant resemblance
to each other. We need scarcely remind our readers that
with regard to the most important and personal progress
which goes on in children, the greater part is effected without our concurrence, often indeed in spite of us. There is
at least very little that we can, or that we know how to,
watch over and direct. But nothing is more interesting than
the study of those points in which we can to some measure
co-operate.

We will take the case of a child of ten months who has
for some time been learning to walk. His first efforts were
very laborious; although carefully held by his frock, he
often failed in his attempts at making steps, and more than
once he rolled over, and the lesson ended with crying.
For a good while he stuck at the A, B, C, of the art, that is
to say, at stamping up and down in one place, like a raw
recruit who is made to stretch out first one leg and then
the other. After a time he became a little firmer on his
thighs and at last he was able to make seven or eight steps
without stumbling. But he still keeps his head constantly
turned towards the person who is holding him up. He
knows that it is only thanks to this support that he is able
to keep upright ; he remembers his numerous falls, or perhaps, like a cat set down for the first time on a slippery
floor, he has an instinctive feeling of the difficulty of the
enterprise ; whatever, the reason, however, he often exhibits
fear. But, encouraged by his increasing successes, he
finally forgets his apprehensions, grows eager and inspirited,
and executes a few well-formed steps, more or less firmly,

and with evident delight. I am even inclined to believe that something akin to pride,—to the exultant feeling of a difficulty conquered,—has been awakened in him. And this sentiment must be experienced in a more or less exaggerated degree, for he measures the importance of his efforts by the trouble they cost him ; and the distance traversed by his footsteps is estimated by comparison of the surrounding objects with his own dimensions.

The eminent philosopher, Herbert Spencer, has admirably explained these two psychological facts. On the one hand he says :—

"The sense of effort which a child experiences in raising a weight, greatly exceeds in intensity the sense of effort it will experience in raising the same weight by the same muscles twenty years afterwards. At maturity a like amount of sensation is the correlate of an increased amount of produced motion. Similarly this relation varies quantitatively as the constitutional state varies. After a prostrating illness, the feeling of strain that accompanies the raising of a limb, is as great as that which in health accompanies a considerable feat of strength.[1]

"The dimensions of our bodies and the spaces moved through by our limbs, serve us as standards of comparison with environing dimensions ; and conceptions of smallness or largeness result according as these environing dimensions are much less or much greater than the organic dimensions. Hence, the consciousness of a given relation of two positions in space, must vary quantitatively with variation of bodily bulk. Clearly, a mouse, which has to run many times its own length to traverse the space which a man traverses at a stride, cannot have the same conception of this space as a man.

Quantitative changes in these compound relations of co-existence are traceable by each person in his own mental history, from childhood to maturity. Distances which seemed great to the boy, seem moderate to the man ;

---

Herbert Spencer, *Principles of Psychology*, p. 204.

and buildings once thought imposing in height and mass, dwindle into insignificance." [1]

It is the same with all the conscious efforts that a child makes, and which habit so quickly transforms into reflex actions.  I suppose that the child of whom I shall have occasion to speak later on, who seized hold of two feeding-bottles at once and lifted them without help, thought he was lifting two objects of enormous weight.   In like manner, when at three-and-a-half months children begin to be a little less awkward in their way of feeling and holding things, it is evident that joy in overcoming diffi-culties is combined with the pleasure they experience in touching and handling.  At this age the mere fact of hold-ing in the hands a coveted piece of paper, of hearing it crackle in the fingers, seems a wonderful feat.   But one of the triumphs that a child enjoys the most keenly, is when he is first able, however faintly, to imitate some of the words which his attendants have repeated to him with so much patience, and which, owing to the quick impression-ability of his brain, had been promptly fixed in the memory, but which the rebellious vocal organs had long refused to utter.   And when at last he succeeds in pronouncing them, however badly, he makes up for lost time, and seems to delight in deafening himself incessantly with the repetition of these sounds.

## III.

### MOTOR ACTIVITY AT FIFTEEN MONTHS.

At a year old, children begin to toddle a few steps by themselves, letting themselves go from one person to another; they are now no longer so much afraid of falls, which they can ward off by reaching out their hands to the floor; they take kindly, moreover, to the *rôle* of quadruped, which they find very convenient.   It is curious to watch them at this age, leaning with their stomachs against chairs or benches, as they will for half-an-hour at a time, arranging and dis-arranging their toys; spreading out little mimic feasts, with

---

[1] Herbert Spencer, *Principles of Psychology*, p. 213.

any scraps they can get hold of ; playing with the cat or the dog, who give themselves up most amiably to being tortured ; then suddenly turning round to any one present ; stretching out their arms with earnest pantomine, if the person moves away or refuses them what they pointed to ; turning this way and that, as their mobile impressions prompt them ; creeping carefully round a chair or table, clinging on to it all the time, stooping down with the same precaution in order to sit on the ground, and then lifting themselves up again with a little more difficulty—in short, learning to trust to their own resources, and now almost able to stand upright and move about at will, with very little help from grown-up people.

Towards the age of fifteen months children will execute with a skill and precision comparatively great, a number of movements which they have either just acquired or have been gradually bringing to perfection. A few instances that I shall cite will give an idea of the enormous progress accomplished by this time. The head can now be raised or lowered, turned to right or left, held immovable by the tension of the neck, shaken to signify *No*, nodded up and down for *Yes*, tossed about with joy, or, at some outburst of tenderness, sunk in the shoulders, or hidden playfully in the hands.

The ear and the eye have become accommodated to distances ; the ear is now always promptly turned towards the point whence a sound is heard to come ; it can also hear more sounds, and has even the power of choosing what it will listen to, and of shutting itself against sounds which are displeasing to it ; it has further learnt to distinguish many creatures and objects by the different sounds they produce. The eye has acquired a large store of adaptations. Cheselden's blind youth, if such a metaphysical blind being really existed, is already far advanced in the double and reciprocal education of sight, touch, and muscle. This eye has no longer the vacant expression of former days ; its look seems sometimes to go through one, it moves with electric rapidity in response to all outward impressions, of whatever nature ; it expresses with force and delicacy all the various shades of thought, sentiment, and will; it is conscious of what it is expressing, and from time to time it does so with intention.

There is something grand and dignified, so to say, in the happy astonishment called forth in the eye by the discovery of a new fact, and in the confiding and steady attention which it pays to the slightest words, gestures, or looks of a person who is speaking. Laughter and tears, which are as frequent as a few months since, are now more often and more fully expressive—intentionally so also—though they do not always express the sentiments, and above all the shades of sentiment, which they are intended to do later on.

As for the hand, the human organ *par excellence*, the stages of its progress must of necessity escape the analysis of an observer attempting to record them; for the movements which it executes, nearly all of them complicated, and most delicately combined, are the result of efforts and acquirements, and degrees of perfection which have gone on from hour to hour during long months. At fifteen months the hand can already touch with more or less certain discrimination and appreciation; it can sometimes measure the effort required by the nature of the difficulty, either known or inferred; the fingers, always in motion, often double themselves up to distinguish the roughness or smoothness of objects, or to find out whether they are hot by skimming the surface. The fist no longer closes with automatic indifference; it now expresses anger, shows an intention to strike, beat, or thump. The first finger often starts out by itself, and is stretched forward to point out or name objects; or the fingers will open out and the hand be waved gracefully to make a salute, or energetically to repulse anything that annoys. In short, the hand can now hold, lift up, and carry weights adapted to the strength and necessities of the child; it is master of the playthings which are its owner's treasure, and, what is neither the easiest nor the least valuable step in advance, it can carry a spoon and a glass more or less skilfully to its mouth.

# CHAPTER III.

SENSATIONS may be studied from three points of view,—the perceptive, the emotional, and the instinctive; that is to say, in the perceptions or ideas which they leave behind them, in the pleasant or painful emotions which they occasion, and in the tendencies or inclinations which they engender. We will begin by showing the order of the natural development of perceptions in a child, as far, that is, as it is possible to observe or to infer it.

"The mind of man," says Bacon, "must work upon stuff." Sensations are the primary matter of the mind, the determining cause of our ideas. But we cannot affirm on the evidence of facts that intellectual operations are, like ideas, engendered by the medium of sensation. If we regard sensations as a particular state of the sensory centres, ideas as special modifications of the intellectual centre. and the various operations as special dispositions of the spiritual organs, the one and the other tending from habit and practice to reproduce themselves, and to excite each other mutually, it remains to be proved, with the help of the scalpel and the microscope, that the transformation, for instance, of the idea into attention, produces a new and persistent disposition of molecules in the tissues and cells of the brain. No observations of this nature have yet been made. But there is no longer any doubt in the mind of anatomists and physiologists, that the play of our faculties is intimately connected with the perfection of the instrument. "Not only does the slightest pathological modification affect the whole of the cerebral functions, but we find these functions increase as age multiplies the cerebral fibres and cells, or complicates their relations to each other. Science has not yet arrived at specifying the functions of each fibre in this wonderful labyrinth ; they count by thousands ; most of them measure less than the thousandth part of a milli- metre ; some of them even can scarcely be seen through the best microscopes." [1] It is the condition of this instrument in the new-born child that we want to understand.

Comparative psychology and physiology, as I have already said, can hardly give us any information concerning the state of the centres of perception during the first period of life. Brain anatomists, however, foresee the moment when experiment may open this vast field to human thought. "It would be extremely interesting," says Ferrier, "to ascertain whether, in an individual born blind, the sight centre presents any peculiarities, either as regards the form of the cells or their processes, or otherwise, differing from those of the normal brain. If such were detectable, we should come near arriving at the characters of the physical basis of an idea." [2]   Dr. Tarchanoff has made some experi- ments on the nerve centres, and especially on the psycho- motor centres of new-born animals, and their development under different conditions, the interesting results of which I will give here in his own words, "In a new-born rabbit the auditory passage of the ear is closed, and does not begin to open till the fifth day after birth, when it has the appearance of a very fine slit. The first sign of a slit between the eyelids appears towards the tenth or eleventh day ; and towards the twelfth day, in most cases, the eyes are quite open. The psychomotor centres generally make their appearance towards the twelfth or thirteenth day.   It

---

[1] G. Pouchet, *Analyse du Livre de M. Ribot sur l'Hérédité*, dans le Journal *Le Siècle*, juin, 1873.
[2] David Ferrier, *The Functions of the Brain*, p. 260.

is the motor centres of the jaw—the centres **of** mastication —which develop first on the grey cortex **of the** brain. After these, the motor centres of the anterior paws appear, and three or four days later, those of the posterior limbs. By the sixteenth day, all the psychomotor centres of the rabbit are fully developed. An almost identical **order of** development of the psychomotor centres has **been described** by Soltman in the dog."

What knowledge have **we of the** condition **of the seat** of perception or **ideation, whether visual, auditory, or** even motor, in new-born infants? Almost none. **In the** first place, as far as vision is concerned, although the constituent parts of the ocular apparatus **may be** sufficiently developed, and though the child sometimes shuts its eyes under the action of a bright light, as it is continually shutting them, it is difficult to draw any inferences from this as regards sight. During the first days of existence, **the** immobility of the pupils and of the iris appears **to indicate** insensibility of the retina to light. It is however probable that it soon begins to distinguish feebly between day and night. This stage is generally reached by the end **of the** first week. A few days after, the eyes begin to follow the direction of light, and of candles. But, except when the child is sucking, the eyes continually wander; they do not fix themselves on any objects or follow their movements, which is a proof that they do **not distinguish** them better than at the moment of birth. **They do not** begin **to** distinguish objects till the end **of the fourth** week, **and then** still very confusedly.

We know that new-born babies **are deaf, the external** auditory passage being closed, **and the** cavity **of** the tympanum containing too little **air.** But we are ignorant as to whether they arrive at the faculty of hearing by intermediate degrees, or by a sudden bound, their auditory apparatus becoming all at once sufficiently developed to fulfil its functions. It is easy, however, to observe during the first fortnight a very great susceptibility to the slightest sound of any kind. A baby shudders and blinks its eyes when it hears any sudden noise, a door being closed, a

piece of furniture moved, the rolling of a carriage, a sneeze,
a burst of laughter, a scream, or a loud song.

As regards muscular perceptions, which, as we have said,
are no doubt faintly begun in the fœtus, their progress in
number and differentiation is very limited during the first
month. Muscular perceptions, being connected with
motivity, are still very rudimentary at this period; but this
does not apply to the other muscular sensations, or to
cutaneous sensations. Young infants, being continually
subject to new sensations of contact, pressure, and temperature, are able to retain and develop sufficient perceptions, by the laws of integration and differentiation, for
the faculty of localizing sensations from external causes
in the different parts of the body to have made some progress at an early period. But what point has been reached
by the second month in the knowledge of external objects?
And, first of all, how does a child see them at this age?

It would be well, perhaps, to ask oneself whether a little
baby a month, or even two months, old sees all the objects
situated in the field of vision. Is there for the individual
also a progressive evolution of the sense of colour, as
Gladstone and Magnus assert that there is for the race?
According to their theory, the sense of colour has only
been developed in man since the heroic ages, i.e., about
3,000 years. The ancients before Homer's time only distinguished light as brightness and colour. As the education
of the organ progressed, three principal colours seem to
have been apprehended by it, and in the order of greater
or less refrangibility, viz., red, green, and violet. In the
second phase of development, the sense of colour becomes
quite distinct from the sense of light. Red and yellow,
with their various shades, including orange, are now clearly
distinguished. The characteristic of the third period is the
power of distinguishing colours which, as regards brilliancy,
belong to neither extreme but are on the whole varieties of
green. Finally, in the fourth period man begins to distinguish blue. This phase is still going on, and with regard
to some portions of humanity is not yet far advanced; we
ourselves, indeed, easily confuse blue and green by candle-

light.[1] M. G. Atlen, as the advocate of evolution pure and simple, has undertaken the refutation of the theory of historic evolution; but there remains still the ante-historic question. The formation of the human visual organ as it exists at the present day, and the development of the sense of colour, can only be explained by hereditary transmission. Our more immediate ancestors of the pre-historic age, did they see the same colours that we do? Do children see colours in the same way as we ourselves? Certain pathological facts engendering optical illusions, and in particular, Daltonism, which is so frequent in adults, might lead us to suppose, considering the incompleteness of the visual organ and the optic centres in new-born children, that they do not see all colours at the beginning of life, that the sense of certain colours is perhaps wanting in them at first, and that these deficiencies in sight vary with different individuals, and in each individual according to the physiological state of the organs, and even according to the day and the hour.

But what data have we for establishing this hypothesis? I believe, moreover, that the value of chromatic distinction is of little importance to the question in point. The non-distinction of colour does not necessarily imply the non-perception of light; and it is by their degrees of light and shade, by the greater or less intensity of the impressions they produce, that a child distinguishes objects.

It is probable that the field of vision only opens out gradually to a child, and that the different parts of it are only apprehended by him according to the degree of intensity of their light or colouring. We should form a wrong idea of the first perceptions of children if we suppose them to be like Cheselden's blind youth, to whom different objects placed before his eyes seemed only a mass of colours spread over a plane surface. Binocular vision, as soon as it operates with regularity of adjustment, produces the perception of coloured space in two dimensions, and suggests the idea of the third dimension. This power is increased

[1] See the interesting review of M. G. Atlen's book by M. A. Espinas, *Revue Philosophique*, Jan., 1880.

in the child by the simultaneous seeing and touching of things brought near to it, but the progress is very slow up to the age of six weeks.   Towards the end of the second month, however, a marked improvement begins.   First, the child pays a little more attention to the impressions which come to him.   His tactile and muscular functions (the latter especially) proceeding from centres more fully developed, bring their contingent of perceptions indispensable to those of sight.   The eye, moreover, has grown more mobile, its muscles are stronger, and it opens wider, thus not only giving an enlarged field of vision, but also allowing the formation of ideas of localization, relief, and distance.

That all these ideas are still very confused in the brain of an infant of two months, is however a matter of course ; but it will already have had a great number of those muscular experiences on which the formation of all ideas of outside things, and above all the distinction between the external and the internal depends.   Just as by the distinction between his own cries and the voices of others, a child at once conceives as distinct from himself other beings capable of making themselves heard like himself, so the muscular sensations produced by movements which he makes himself are distinguished by him from the sensations resulting from movements foreign to himself, and this distinction is corroborated by the concomitant sensations of sight, which cause him to see foreign bodies in motion.   His progress in mobility thus carries with it progress in his ideas of the separate existence of things, as well as their forms, their relations to each other, and their distances.   By the age of three months the mobility of his eyes, neck, and arms has increased, and hence a quantity of muscular sensations combined with visual ones, which result in a clearer distinction of all the ideas of which we are speaking.   He also begins now to have the power of discerning,—I do not say of appreciating,—weight, which calls into play the muscular sense of effort, and above all, if we go by Ferrier, the feeling of the contraction of the respiratory organs.[1]

---

[1] Ferrier, *The Functions of the Brain*, p. 358.

Mary, at the age of three-and-a-half months, can already distinguish several parts of her body. When her mother says to her, " Where are your feet ? " her eyes first of all move uncertainly to right and left, and then, bending her neck forward, she directs them towards her feet. She plays with and fondles her mother ; when her mother's face is bent over hers, she seizes it with her little fat hands, and touches and pats it with an evident intention of showing tenderness. "She chatters to the flowers," to quote her mother's words. She is passionately fond of colours, especially very brilliant ones. If a coloured picture is shown her, she makes two or three sudden starts, and then holds her little hands out eagerly towards it, palpitating with desire or pleasure, her eyes attentively fixed, her face beaming with joy, and uttering all the time little birdlike cries. But seeing is not enough ; she soon wants to handle the beautiful object, and seizing it with both her hands she crumples it up and gazes at it admiringly, but without seeing anything more in it than pretty colours. The word *picture* will now make her smile from association of ideas and feelings. She is very fond of babbling to the birds, with whom she is well acquainted ; and she not only turns towards the cage when the canary is singing, but also when he is quite quiet, if her mother says to her, " Where is Coco ? Listen to Coco."

She understands, from the expression of the face and the tone of the voice, when she is being scolded, and then she wrinkles up her forehead, her lips twitch convulsively and pout for two or three seconds, her eyes fill with tears, and she is on the point of sobbing. She is very sensitive to caresses, and laughs and plays with every one who laughs and plays with her. But she is jealous in the extreme. If her elder sister is placed beside her on her mother's lap, and the mother kisses the sister, quite a tragic scene ensues ; for a few seconds she remains still with her eyes fixed, then her mouth begins to twitch, her eyes fill with tears. She becomes convulsed with sobs, she turns away her head so as not to see her rival, and presents for some minutes the picture of misery. She behaves precisely in the same manner if her mother gives her sister the feeding-bottle, or if the latter

takes it up from the table ; but when the mother takes the
bottle herself and pretends to suck from it, she instantly re-
covers her serenity : it seems as if she had no sense of egotism
in her relations with her mother.

Let us take another child of different sex and temperament. Georgie is seven months old. He has hardly entered
my room when his attention is vividly excited by the noisy
movements of a sparrow hopping up and down in its cage
close to the window. Then he fixes his eyes steadily for
the space of three minutes on a cat lying cuddled up at the
foot of an arm-chair : he has often seen cats before.

But the sparrow sings a few notes, and Georgie looks
round on all sides, not knowing whence the pleasant sounds
proceed. I call him by his name, Georgie ; and though he
has never heard my voice before, he smiles at me most
sweetly. The next thing that attracts his attention is a
bunch of flowers which I placed on a table near him. He
stretches out his arms towards these, and evidently derives
great pleasure from looking at them ; but his delight does not
manifest itself in those bounds and cries and outbursts of joy
which I have already noticed in little Mary and many other
children of the same age (three-and-a-half months) under
similar circumstances. Georgie is a big, fat boy of Alsatian
parentage, chubby-faced, grave, slow, and obstinate ; while
Mary, on the other hand, is a pale, slender, animated, sparkling little Parisian. Ten days after his first visit to me,
Georgie was brought into my room for a second "sitting."
This time he amused himself with making jumps at my cat.
When tired of this game he stretched himself forward to seize
a plate which stands in the middle of the table ; I allowed
him to reach across the table and to touch the plate ; he
handled it with evident gestures of pleasure, and his face
also expressed extreme delight. After a while (I supporting
three-fourths of the weight) he carried it to his mouth, as he
does with all objects he gets hold of. His grandmother has
brought him up with a bottle. I have noticed in this child,
as in many others, a quite peculiar tenderness for his grandmother, who has been his nurse. The instant he has got
hold of a desired object, and whenever he experiences any

pleasure which is not connected with eating, he turns smil-
ingly towards her, as if his joy were not complete till he
had shared it with her. Or must we see in this nothing
more than a mechanical habit, the child having never experienced any joy without his grandmother being at hand to
take part in it?

Thus in children of three-and-a-half and seven months we
find the power of distinguishing a large number of essential
ideas. They can distinguish themselves from their mother,
their grandmother, their sister, myself, a bird, a cage, a cat, a
table, a plate, etc. But even at seven months they see prominent details better than whole objects; they see them by a
kind of process of *imaginative abstraction*, by means of which
external perceptions come to them as scraps of colour.
They have only an imperfect appreciation of distance and
weight; an object a little way off must be very bright or
very sonorous, for their curiosity, even when excited by
craving, to fasten upon it.

They try to seize objects before they are within their
reach; they will put out a whole hand and all their strength
to take up a bit of coloured paper, as if it were a compact
and heavy substance. At three months a child can distinguish his feeding-bottle by its form and colour; but so little
does he compare, that he will seize an empty or a full bottle
with equal eagerness. At seven months he compares better
than at three; and he appears at this age to have visual
perceptions associated with ideas of *kind;* for instance, he
connects the different flavours of a piece of bread, of a cake,
of fruit, with their different forms and colours. At one or
two years old, when his curiosity has been developed by
exercise, by the continual supply of fresh emotions, by the
growth of his muscular forces, and above all of his mobile
and locomotive faculties, his power of comparison is very
much greater. All his ideas of situation, of figure, of relief,
of distance, and of weight have now reached such a pitch of
perfection that in many cases they are nearly equal to those
of an adult. But many are the illusions he will suffer, sometimes consciously, from these perceptions, which bring into
conflict the actual sensations of sight and the judgment ac-

quired from outside. Countless experiences, countless errors, deceptions vividly felt, will have to be gone through before the child can arrive at the summary knowledge of the external world, such as it appears to the eye of an adult. This knowledge, in point of fact, has no limits ; or rather, its limits are for ever receding before science.

## II.

### THE FIRST PERCEPTIONS.

*The Sensations of Taste.*—The first manifestations of pleasure in little children are connected with the sense of taste. A few hours after birth, hunger makes itself known by the efforts of the mouth, as if to seek its food, and its attempts to suck any object presented to it, and by wailings accompanied with lively movements of the arms. During the following days, when the infant has learnt to suck, it will remain glued to the breast, and, until its appetite is satisfied, the strongest appeal to its attention will not disturb it. There is no doubt, however, that after the child has sucked for a few seconds it goes on doing it mechanically, without any feeling of pleasure or otherwise, except at intervals, its organs being for the present incapable of feeling a sensation repeated too often. The sensation of taste soon ceases to react on the nerves subject to its influence, and after a certain time the intervention of a stronger excitement is needed to produce a reaction. Greediness then comes in as a stimulus, and in a child is as legitimate as it is common.

We are at liberty to believe that the sense of taste is very slightly developed in a child just born, the indirect proof of this being the small necessity there is for a child to distinguish flavours. But we have more positive reasons also for this assumption ; we know that in adults the sensations of taste are mixed up with sensations of smell which influence us in distinguishing flavours. " If our nostrils were closed," says Longet, " we should not be able to distinguish vanilla cream from coffee cream, and both would only produce a general sensation of softness and sweetness." But also, as

Brillat-Savarin has told us, " the empire of flavour has its blind and its deaf, and degrees of sapid sensitiveness, very different in different people, are seen already at a tender age." In some cases children of six months have been induced to take disagreeable medicine simply by a change being made in the colour. Others again, at an earlier age, will refuse to suck from their mother, or from particular nurses, because of something unpleasant either to their sense of smell or taste. Some children are very easily disgusted. I have seen a child two months and a half old refuse its bottle determinately and with a face of disgust, once because it was filled with water, and another time because the milk was not sweetened. Some children appear sensitive to all impressions of taste, whatever they may be, while others again are indifferent to them. A member of my family, when a child, could never be persuaded by her mother to taste wine ; and, in spite of the advice of doctors, she has never been able to drink anything but water. Other children, and by far the greater number, begin very early to notice the acid taste in certain substances. In general, however, children very easily change their tastes, which is a reason for not forcing them to eat things against their inclination when there is no necessity for it.

*Olfactory Sensations.*—Children, as a rule, appear to remain for some time insensible to bad smells. The probability, however, is, that they are only less sensitive to them than adults are, and that their olfactory apparatus, the delicate organs of which are closely related to the different regions of the brain, are not highly developed in the earlier months. This would not be at all surprising, as the olfactory sense seems to be of no use whatever to the nursing child. Possibly also, as odours are variable and fleeting in their nature, it may require a practised judgment to distinguish the sensations produced by them, and to refer them to the right objects and causes. To know the origin of a sensation, is to be able to single it out from the concomitant sensations.

Nevertheless certain specialist doctors have assured me that new-born infants are impressed by smells, and I have

D

had cited to me the cases of a child of six weeks, and
another of two months, who, in refusing or in taking the
breast of certain nurses, were guided only by the smell of their
perspiration.   Tiedemann, whose son at the age of thirteen
days had rejected different medicines after having tasted them
several times, supposes that "he distinguished them from
his food by their smell."   A son of Darwin's thirty-two days
old, "recognised its mother's breast at a distance of 75 to 100
millimetres, as the movement of its lips and the fixity of
its eyes testified;" and Darwin assumes that sight and touch
had nothing to do with this, and that the child was guided
by the sensation of heat, or by the smell.   But scientific observations concerning smell, both from the emotional and
the cognitive point of view, are very incomplete, even when
coming from the most competent people.

An infant of fifteen days, one month, or two months,
*manifests* only visual or tactile sensations in the presence
of, or in contact with, a rose, a lily, a geranium, or a nosegay of flowers ; but I would not affirm that it *experiences* no
other sensations.   I have subjected a certain number of
children between the ages of ten and fifteen months, to
experiments relating to the olfactory sensations, and all of
them,—excepting one who was insensible to any smells, even
those of tobacco and ether,—felt the different olfactory impressions very vividly.   One child of ten months seemed
to me to be very sensitive to pleasant smells and very
much annoyed by bad ones.   When I prevented his seizing
hold of a rose, or a bunch of violets, which I had held close
to his nose, his expression and his gestures evidently begged
me to give them to him.   When I again held them up to
his nose, he remained some time quite motionless, smiling
with pleasure ; in short, he seemed to appreciate and delight
in pleasant scents.   As we said above, there are some kinds
of food which not only impress the organs of taste (affording
us the sensations described as sweet, saline, sour, acid, bitter,
etc.), but which also act on the olfactory nerves,—chocolate
and coffee, for instance,—and these this child was extrava-
gantly fond of ; and it certainly appreciated, as thoroughly as
a grown-up person could, the perfumes of cocoa and mocha.

I have seen two other children of the same age who cared less for the perfume of a rose or of mignonette, by which the alimentary organs are not affected, than they did for the smell of chocolate. Odours connected with food are apt, as we can easily conceive, to take rank before odours pure and simple with the inexperienced sensibility of a child, while the contrary is frequently the case with adults. The ancients crowned themselves with roses at their banquets; and we ourselves often deck our dinner-tables with flowers, or at any rate we generally have about us delicate perfumes which do not impair the flavour of the wines and viands.

*Organic Sensations.*—Bain has given this name to those sensations which arise from the diverse dispositions and affections of our bodies and which all become fused into the vital or fundamental sense—the general sensation of existence and the immediate sensation of our own body, which is, according to Luys, the sum of all nervous actions, and, as it were, the organic basis of consciousness. To the organic sense are to be referred the respiratory sensations, which we have already spoken of, the sensations of the circulation and nutrition, the organic sensations of the nerves and the muscles, such as fatigue, pain from a cut or a scratch, cramps, spasms, etc. There is no doubt that children, with their incomplete organisms and imperfect powers of adjustment, frequently suffer from these different kinds of pain, especially during the first period of existence. It is less evident that they experience corresponding sensations of pleasure, the latter being only equivalent to a passive state of functional regularity and general well-being which is not expressed by violent signs.

The state of comfort resulting from regular breathing is one of these passively pleasurable sensations which children have most experience of. In fact, respiration being more active in infancy than in adult life, and the frequency of inspiration in children causing them greater need of oxygen, they must on the one hand have their attention a good deal excited by this respiratory action, on account of the rapid modifications which it causes in their organs ; and, on the other hand, the quality and temperature of the air breathed must afford them sensations corresponding to the freshness and appetition of their respiratory organs. We know, moreover, that even before the age of a month there is nothing more salutary or enjoyable to infants than to be carried out into the open air.

If we admit that there is an organic and automatic consciousness always watching over all the internal and external parts of a young infant, we may also class among its happy sensations those which regular sleep afford it. Sound slumber in a child implies healthy and thorough nutrition, and perfect harmony of all its functions. Not only is sleep the necessary condition of the growth of the organs, and of moral and intellectual development, which depends less on the impressions received than on the perceptions digested and consolidated, but it is the greatest boon that nature bestows on children before the age of eager curiosity and easy locomotion.

*Muscular Sensations.*—I have already had occasion to speak of a sense to which a certain number of the tactile properties of bodies have been transferred—the muscular sense ; which, in the opinion of Renouvier, is only an hypothesis, but the existence of which Bain and Wundt believe they have established by irrefutable experiments. To this " interior contact " are usually referred all sensations of pressure, weight, traction, and even resistance; it is regarded as the conscious centre, not only of muscular contractions, but even of the state of the muscles, and, in a lesser degree, of the state of the articulations of the skin accompanying muscular contractions.

Whatever be the origin and whatever the centre of these sensations, it cannot be denied that there result from them sensations of pleasure or pain of a particular sort, independent of the moral pleasure or displeasure resulting from desires more or less keen and more or less perfectly satisfied. An effort pleases in itself when it is not too violent ; it may then be said that sensations of pressure, of weight, of muscular efforts, are agreeable to children when not beyond their strength, and more or less disagreeable to them in the

contrary case. These sensations are also all the more
agreeable the more they are varied. Each limb, and each
organ is capable of an infinite number of movements which
produce in the owner the personal, although perhaps unconscious, sense of his own activity and existence. How
many muscles are put into play successively or simultaneously by the simple action of dragging along a ball tied to
a piece of string! And in the matter of a weight, however
small, to be lifted or held up, it is not only the muscles of
the back, shoulders, neck, or arm which are exerted, but also
a certain number of respiratory muscles. Ferrier seems in
fact to have demonstrated that the sense of effort necessary
for the discernment of a heavy weight "would be more
correctly assigned to the region of the respiratory muscles."
This sensation of weight, which precedes the appreciation
of weight, is thus a muscular phenomenon of such complexity
that the host of sensations which result from it must deeply
affect the muscular consciousness, or the centre where the
attributes of that consciousness are located. The pleasure
of a state of equilibrium and health, the pleasure of moderate
and appropriate exercise when he moves his own limbs or
they are moved for him, such are the enjoyable sensations
which unconscious infants experience every day. But on the
other hand, how many disagreeable and hurtful sensations
do not we ourselves cause them, without being aware of it.

*Thermal Sensations.*—The sensations of heat and cold
depend essentially on the difference of temperature between
our organs and the surrounding atmosphere, the radiating
body and the body in contact—either the air or any object
whatever. When the disproportion is only moderate, there
results an agreeable sensation of warmth or coolness, as the
case may be ; when it exceeds certain limits, it causes pain
and at the same time more or less serious trouble in the
depths of organic life. In spite of the incomplete development of their " plexus " and nerve centres, children, from the
delicacy of their epidermic tissues, their slight degree of
nutritive activity, and the smallness of their bulk, are predisposed to very great susceptibility to temperature. We
have already seen that the fœtus is very sensitive to high or

low temperature.   With all new-born animals, the tendency
to suffer from cold is very great, and they are very apt to
die of cold even in summer, if out of reach of the sun.

The pleasure which young children derive from sensations of warmth is so evident that it is needless to dwell on
it.   But we may legitimately ask whether the sensations of
cold which cause death to so many are as painful to them
as they would be to older children or adults.   My own
opinion is, that very young children suffer less in reality
than we should expect, considering the sensitiveness of their
organisation to such impressions.   Sensations of the above
nature are generally modified in the adult by judgments,
habits, and sentiments of diverse and variable kinds which
a rise or fall of some degrees may awaken in them.   A
sentry on duty during a hard frost will possibly have
thoughts in his mind which may counterbalance the keen
sensations of cold, or may make reflections and comparisons,
call up recollections, indulge in imaginations, which will
have the result of making the cold more painful to him than
to a child of two months, or only a few weeks, exposed
perhaps on a doorstep a few paces from him.   It is needless
to insist on a fact patent to all, viz., that the individual
constitution, and the accidental state of health augment or
diminish, in children as well as in adults, the susceptibility
of which we are speaking.

*Sensations of Touch.*—The sense of touch is a means of
preservation before it becomes a means of instruction ; and
we know that, like the muscular and thermal senses, it
operates already in the womb under the action of the
pressure of a hand or of heat or cold air.   At the moment
of birth cutaneous sensibility is very acute, and the child
seems to show, either by a happy state of quiescence or by
cries and convulsions, that the contact with outward objects
causes it pleasure or pain.

The sensation of resistance is considered by some to be
the fundamental sensation of touch.   The intensity of this
sensation, and the modifications of which it is capable,
according to the nature of the objects which cause the
impressions, will produce in children feelings that are either

really painful or simply annoying and irritating; such are
the feelings of tickling, rubbing, bandaging, pricking, etc.
Sensations of contact which seem to us insignificant cause
a child great uneasiness, and will cause it to make faces,
to scream, to wriggle its body, to toss its arms about and
carry them automatically to its face. A feather passed
over the eyes and nose of a child fifteen days old made
it frown, contract its nose obliquely, and close its eyes.
Other children again, at an older age, are insensible to this
kind of excitation.

As to the pleasant sensations of touch, it is not so easy
to tabulate their signs (though, as we have said, they are
very evident) in children less than two months old. Light
and delicate pressure, the contact of a soft skin or material,
do not call up in the face any decided expression of
pleasure, a smile or movement of the eyes. Infants, however, are not insensible to this kind of sensations, which
signify to them already the presence of certain objects; it
is impossible but that they should produce in the little
creatures some vague sensations of comfort, in spite of their
inability to localize or differentiate them. Later on, experience and comparison will have taught them to distinguish
these sensations from the more violent ones which cause
pain or annoyance. But at the age of two months, when a
child appears to enjoy the touch of my hand stroking its
own hand, or cheek, or forehead, is it the contact itself
which pleases it, and even makes it smile, or is it the idea
of its mother's breast which the contact of the skin calls up,
or the pleasure resulting from the sensation of temperature?
Possibly all these at the same time. Soon too the sympathetic significance of caresses, which babies are not slow
in understanding, begins to afford them very evident pleasure
in connection with tactile sensations. However this may
be, and notwithstanding the absence of signs indicating it,
I incline to believe that tactile pleasure is experienced at a
very early period. There is no doubt about it in the case
of young animals. If I pass my finger, gently, and several
times in succession, over the head of a sparrow ten days
old, it will almost close its eyes, and with its head bent

down it will put itself into a position favourable to the continuance of this pleasant process. The same thing happens with kittens and puppies. And why should it not be the same also with young children?

*Visual Sensations.*—We cannot but regard as exaggeration Tiedemann's statement concerning his son, who, when scarcely thirteen days old, he tells us, showed, both in his eyes and his general countenance, expressions of pain or pleasure at the sight of certain objects, and paid sustained attention to the gestures of people who spoke to him. But to keep within well-established limits, we may without audacity affirm that at the age of a month or forty days a child has already experienced a certain number of pleasures and pains suggested by visual impressions. The pleasures consist in the sensations caused by luminous objects—candles, lamps, the flames of the fire, the sunlight, brightly coloured objects, and the lights and shadows caused by moving them about. The pains are caused by impressions of too violent a nature, colours which are too bright, sounds which are too noisy, objects brought too abruptly in contact with the retina, or too rapidly shaken in front of it, and also by the moral annoyance which must result from imperfect adaptation to one's surroundings, or at any rate the physical pain arising from efforts after normal adjustment. This latter hypothesis seems to me indeed to rest on grave analogies. It is unfortunately not true that the progressive adaptation of the young human being to the surroundings for which it is hereditarily constituted takes place by successive steps, with gentle transitions and providential management. As a matter of fact, the sufferings of a human being are all the greater in proportion to his weakness. With regard to intellectual perception, the organs may exercise themselves usefully according to their strength; but the same cannot be said of sensibility relatively to pleasure. As Rousseau has said, "Much time is needed for learning to see," and imperfect sight is necessarily accompanied by painful sensations, like all other ill-satisfied needs.

Visual impressions do not produce in children the same emotions of pain or pleasure, nor perhaps so great a number

of them, as in adults.   All colours, it is true, attract and
fascinate them ; subdued colours, also, are not always indifferent to them, and sometimes they afford them evident
pleasure, provided they are distinct and contrasted with
brighter ones, as black on grey, or even grey on white, and
provided, above all, that the child's organism is impressible.
I saw a little girl of three months and a boy of five months
delighted with some drawings of a uniformly grey colour.
The boy took a particular fancy to some lithographs in
black and white ; it was enough to say the word " picture "
to him, for his eyes to turn to the part of my room where
these lithographs hung.   Another child of six months only
evinced pleasure at the sight of bright-coloured pictures and
flowers; but these seemed to delight him quite as much as
they did the other children.   The conclusion I came to
was, that, owing either to hereditary causes or to personal
habits, there was less energy in his visual organs, or that his
moral sensibility was not so easily excited by the sensation
of colour as that of his two companions.

It would not be very easy to discern in a child between
the ages of one day and five months the painful emotions
produced by certain colours, although it is incontestably
established that the visual organs have adaptive powers,
and also without doubt likes and dislikes.   I have not
succeeded in discerning in children of this age, or even in
much older ones, any trace of those affective predispositions
of sight, the result of differences of organisation, which
express themselves unconsciously in adults in marked
preferences for such and such colours.

The pleasures and pains of sight are possibly in great
measure artificial phenomena with adults.   But with children who go on for a long time without any distinct ideas
all the utility or the hurtfulness of different objects; all
objects whatever are in their eyes merely moving colours,
all of which give them pleasure as soon as they produce impressions on their visual organs.

*Auditory Sensations.*—Infants are very early excited by
sound, by musical instruments or song.   Tiedemann's son
heard the piano played for the first time when he was forty

days old, and he is said to have shown singular delight
and excitement at the sound. A young relation of mine
delighted in hearing singing or playing when only a month
old. At the age of six months he was taken on a visit to
some aunts; and the strong emotion produced in him by
the songs they sang to him was manifest from the brightness of his eyes and the immobility and heightened colour
of his face. The younger one sang to him first, and he
listened to her with evident delight; but when the other
sister joined in, with her more vibrating and more melodious
voice, he instantly turned to her, and an undefinable expression of admiration and surprise came over his countenance. All children are not equally sensitive to melody;
but all clear, ringing sounds, especially if reiterated more
or less rhythmically, seem to amuse them when they do not
strike too violently against the tympanum. I have seen
many infants of two months whom noises of medium intensity,—the sound of a door being shut, a footstep, a voice,
the bark of a dog close by,—did not seem to affect in any
way whatever. But at four or six months almost all babies
like being sung to, and a great many try to warble themselves, from instinct or love of imitating. We may, then,
conclude that there are sounds which are pleasant and
sounds which are unpleasant to children, according as their
vibrations correspond to certain conformations of the
acoustic apparatus, or to certain inherent states of the
personality, or awaken concordant emotions in the mysterious recesses of their hereditary sensorium.

However it may be, children become soonest accustomed
to those very noises which at first, for one reason or another,
impressed their tympanum disagreeably. Startling, piercing,
scraping noises are not so disagreeable to them as to grown
people. They will shudder, it is true, on hearing them, and
sometimes cry if the noise is very loud and near; but few
sounds displease them on moral grounds, or by reason of
the association of ideas which makes noises call up the
idea of objects known to be disagreeable. They are not
more particular about melody, either, than bees, serpents,
monkeys, and other animals, to whom the most rude and

deafening noise, providing it is rhythmical, is as good as music. From the moment that children can hold things in their hands, anything that they produce sound with delights them, and they are evidently as happy as they can be in the execution of these feats. I can never see little children doing their best to deafen themselves and others without being reminded of the monkey musicians of Africa, of which Houzeau writes :—" The noise of these animals is not always the mere accidental result of the play of their organs, it is sometimes produced intentionally; they will make a noise for the love of the noise, and as a means of amusing and exciting themselves. I am not alluding to the ordinary sounds with which tropical forests resound— deafening screams, the crashing of broken branches, the pecking of beaks on the bark of trees, the cracking of nuts in the teeth—which are the natural result of the existence and occupations of the inhabitants of the woods. I mean the sounds which are produced by monkeys with as determined an intention as that of a bell-ringer or a player on the drum. The black chimpanzees of Africa, for instance, will assemble together as many as twenty, thirty, or fifty, and amuse themselves not only with uttering shrieks, but by beating and thumping on dead wood with small sticks held in their hands or feet." [1]

---

[1] Houzeau, *Les Facultés Mentales des Animaux*, etc., t. ii., p. 106.

# CHAPTER IV

IN man as in animals, instinct manifests itself in congenital dispositions to perform certain definite actions under certain special circumstances. These dispositions to produce given actions are the tendencies which make up our natural and hereditary constitution. It is the unconscious experience of our ancestors adapting itself with more or less consciousness or freedom to the individual experience of their descendants. It is necessary to insist on these mysterious impulses of organic and psychic activity, and to set forth their true nature, in order that we should not be tempted to look on animals as mere machines, set in motion once for all at the beginning of life, and on children (with the exception that they have reason in addition) as animals, with analogous, and perhaps rather inferior instincts.

The best treatises of current philosophy, whether intended to awaken in the young generation or to perfect in the more erudite a sense of psychological observation, still represent instinct as having the following characteristics :—ignorance of the end in view, immediate perfection of special actions, infallibility, immobility, and uniformity in these actions. To be quite just, however, we must own that M. Janet does not recognise " in these characteristics absolute and inflexible laws."[1] He admits as a fact of experience that instinct may vary "under the influence of certain circumstances, though within very narrow limits and in exceptional cases ;" in a word, that these modifications are

---

[1] Janet, *Traité Elémentaire de Philosophie*, t. i., p. 37.

nothing else than " an innate power of adaptation of the animal to its surroundings," and that they only occur in very secondary details.

Let us first endeavour to show what part ignorance, or rather unconsciousness, plays in the working of instinct. It is perfectly evident that not one of the movements of the fœtus can be explained by reason, or intelligent will, or experience. They do not proceed from consciousness, considered as an impulsive cause, although they may be supposed to provoke the exercise of consciousness. When, shortly after its birth, we see a little baby feeling after its mother's breast, and co-ordinating the movements of its mouth, head, and neck so as to suck in the milk ; when it combines the actions of the tongue, palate, and pharynx, which co-operate in the process of deglutition ; when, a little later on, it presses its little fingers and fists against the breast, in order to facilitate the passage of the milk, when the combined and harmonious action of all these numerous organs produces respiration ; when the eyelids close if the conjunctiva be touched, or if too intense a light disturbs the retina, or a violent sound shocks the ear ; when irritation of the face, ears, or tongue causes contraction of the muscles, etc., etc., we know that neither experience nor reason have taught the little creature these movements, which are accomplished with a precision far superior to what we find in actions where the will intervenes. Both children and young animals perform all these actions without knowing either their object or the means by which they are executed ; but they are aware of what they are doing, and the more they advance in knowledge the more fully will they be aware of it, and the more they will realize the means to the end, and the end associated with the means. The impulses are arbitrary and unconscious, but the actions have a constant tendency to revert to their original nature, which may often have been conscious and even voluntary.

Let us proceed to consider the immediate perfection of instinctive actions. " Animals," says M. Joly, " generally succeed the first time, without any preliminary attempts and failures.   Birds have no need to study in order to make their nests.   Carnivorous animals feel no hesitation when they find themselves for the first time in the presence of the prey destined by nature for them ; and amongst the herbs of a meadow ruminants will go straight to those adapted to them."   This perfection and infallibility of animal instinct are very much exaggerated.   The assertion is true in general with regard to the principal mechanical functions of life, the most delicate and indispensable organic actions, the execution of which it would have been dangerous to entrust to the will.

Here the impulse of instinct is certain and unerring, "like all the great harmonies of nature," says Houzeau. But in the greater number of other less necessary actions, which are "ruled or at least influenced by the animal itself," we find an instinct "subject to accidental illusions, to the aberrations of the individual, to the idiosyncrasies of species."[1] It would not be uninteresting to quote a few examples from this same author, showing that instinct can sometimes err.

"The large earth-worm, or lob-worm, is very much frightened of the mole ; and whenever it feels the earth moving, it mounts to the surface to escape from this insect-hunter.   Now certain birds, such as the gull and the lapwing, and fishermen who use earth-worms as bait, prowl about on the sands in search of the latter.   The worm, thinking that the shaking of the earth is caused by the approach of a mole, comes out of his subterranean retreat, and falls a prey to the other enemy.   This is a false application of instinct, but it is not correctly speaking a false instinct.   .   .   ."   In the following facts, however, we see error rather than illusion : "Of what use would it be for a hen to retain the instinct of sitting when her eggs are addled, just as if she were able to hatch them ?   A superior guide would make a distinction.   It is true all the same that a hen will sit without eggs, when the proper time arrives,

---

[1] Houzeau, *Etude sur les Facultés Mentales chez les Animaux*, t. i., p. 295.

from which we surmise that repose, the attitude of prostra-
tion, and a partial abstinence are necessary to the dissipa-
tion of an overabundance of organic heat. The hen which
broods over an empty nest, does so from personal necessity
. . . But the instinct of preservation is sometimes faulty.
Thus, little birds mistake the cuckoo for a sparrow-hawk ;
and these same sparrows attack the European goat-sucker
as if they had reason to fear him, while in reality the latter
bird feeds solely on moths and nocturnal insects. . . .
Instinct here goes too far in the idea of protection ; it may
be said to be an excess of precaution. Here however is
an example of a contrary case. The aphides, or plant
insects, do not know that the larvæ of the syrphides are
their mortal enemies. They feel no fear at the sight of
them and will even walk over their bodies. The syrphides,
which feed on aphides, take advantage of this imprudence,
as may easily be imagined.

"Thus the consciousness of danger, and even of an
ordinary and constant danger, may be at fault ; instincts
therefore are not absolute. But the facts in our possession
are certainly very insufficient to determine the important
question at the head of this chapter—Can instinct err ?"

It is probable, but it has not yet been demonstrated, that
the execution of instinctive actions is influenced in a certain
measure by volition. We may infer this particularly from
the fact that animals have the power of suspending their
respiration and exercising a partial or momentary control
(as man himself can) on certain automatic functions. Dogs
and horses suspend their breath, in order to listen better,
when they apprehend some danger. Now, if certain actions
of the automatic life of animals are in part subjected, though
in a very limited degree, to the control of volition, why
should it be a matter of surprise if one day it were demonstrated that all instincts may be equally influenced by
a like control ?

Since, then, instinct may err, and since, up to a certain
point, it is subject to the control of consciousness and
volition, it follows also that it is neither uniform for the
species nor invariable for the individual. Even in the

case of animals, it obeys the law of progress; and this progress is more or less rapid according to the usefulness
which is to result from it to the species, more or less slow
according to the degree which it is to attain in the species,
and no doubt also according to the aptitudes of the individual and the influence of surroundings.

Colts, calves, and young chickens all walk by instinct as
soon as they are born, with an automatic action facilitated
by their organization and which they perfect by exercise
and attention.  Birds fly, or try to fly, as soon as their
wings are strong enough.  And see how usefully heredity
interposes in similar cases ; sheep cannot take care of their
young ones as monkeys and women do, the lambs would
therefore perish if they could not very soon use their legs
and walk.   A child sucks by instinct, and moreover, like
cats, dogs, and lambs, it very quickly learns to perform this
operation with rapidity and certainty.  With walking, however, it is different ; and although the child may have an
instinctive faculty for the operation, it only acquires it
perfectly by dint of countless efforts and at the price of
many tumbles.   Here too we see the care of nature, for if
the organs of locomotion were perfect at starting, while
intellectual and moral experience is wanting, they would be
the cause of dangerous errors and fatal adventures.  Again,
children very early show themselves adepts at the signs,
gestures, or cries which are expressions of emotion or
volition ; but it is not until their psychic faculties have
already reached a degree of development superior to that
of many adult animals that they begin their first attempts at
speech.

Thus, neither with human beings nor with animals is
instinct a uniform and infallible guide.  For children, as for
animals,—though in the former case within limits generally
less restricted,—there is a more or less easy and certain
development of the instinct to be followed ; a child has to
carry on the simultaneous development of all its senses and
faculties—intellectual, volitional, and moral.  One is struck,
it is true, with the slowness of a child's progress in comparison to that of young animals ; but a child of a year old

is in many respects much more advanced than an adult animal will ever be. In these respects there are great differences in children, according to their natural energy, the force of hereditary transmission, and the nature of the circumstances under which their development proceeds. This should never be forgotten by those who are closely connected with young children. Their influence, whether they will or no, is by no means a matter of indifference, whether from a quantitative or a qualitative point of view, to the development of the instincts of their young charges. They would be making a great mistake if they imagined that nature does everything at this early age, in other words, that instincts, the fruit of hereditary experience, like a sort of providential grace, dispenses young infants from all need of effort and personal experience.

We, who are partisans of the doctrine of evolution, must be careful to avoid the error into which the optimist champions of final causes fell through exaggeration in the opposite extreme. We must not sacrifice personal experience to the experience of the race, nor imagine that the apprenticeship of life is nothing more than reminiscence, that the child has only to repeat mechanically the work of its ancestors. As has been very aptly said, " Notwithstanding hereditary transmission of instincts, everything has continually to be done over again, to be begun anew in each new individual; and life is made up, not of a series of easy reminiscences, but of a chain of laborious acquisitions and personal conquests. In short, in the evolution of the race we must not lose sight of individual evolution."

## II.

### SPECIAL INSTINCTS.

Instinctive activity presents a great many different forms in new-born children, or in children several weeks old ; we shall endeavour to make a rapid analysis of the principal of these. And first of all we must distinguish between general tendencies, or simple appetites and instincts, and special tendencies, or complex appetites and instincts.

E

Among the first we class the instincts of taste, odour, sight, hearing, touch, muscular activity, and thermal equilibration; and among the second the instincts of nutrition, of sleep, of utterance, locomotion, and sexuality.

*General Instincts.*—Wherever there are organs of sensibility, there is also an instinctive tendency to perform actions which awaken a certain sort of sensibility. All the senses desire to be satisfied; and we have seen that from the moment of birth, or at any rate, a few days after, a young child will begin to use the energy which belongs to each of its different senses. This is especially the case with the sense of taste, which nevertheless at the outset is passive and obtuse, because it is confounded with the sense of nutrition. By the fourth month, however, the child has become better able to abstract his various sensations, and shows evident signs of a tendency to desire the food he has recognised as agreeable. It is the same also with the sense of sight. By the third week the eye begins to be attracted by light, and soon after by any luminous or coloured objects near at hand; and we can here plainly see that the instinct which impels every organ to maintain its vitality is in harmony with the instinct which drives them to seek satisfaction in exercise; all colours are pleasing to young children if they stimulate their sight without disturbing it. The same may be said of auditory impressions; we have seen how much young children delight in noise, and how indifferent they are to discord, provided their ears are not shocked. These two senses, sight and hearing, which later on will become two essential instruments of instruction and emotion, operate at first solely for themselves, if I may so express myself, and with an aim entirely affective. A child of two or three months spends the greater part of his waking hours in looking and listening for the sole pleasure that the sensations of seeing and hearing afford him, independently of any immediate or future utility, and without any feeling of curiosity properly so called. The next sense that is awakened is that of touch. At the age of three months, children begin to stretch out their hands in order to take hold of things ; they touch and feel everything within reach, and their tactile sensations develop day by day.

A child of six months spends at least half of his waking hours in exercising the organs which afford him pleasant sensations,—visual, auditory, tactile, and muscular ; and there is no doubt that these exercises have as a rule no other object than the development of the organs, though at intervals the child notices them with attention, and this attention contributes usefully to his progress. But the tendency to enjoy pleasant sensations, and to repeat them over and over again for their own sake, is the dominant instinct at this period. This tendency, moreover, exists at all periods of life. A grown-up person, who is not compelled either by the necessities of existence, or the claims of duty, or the influence of special or professional habits, to bring all his faculties under the discipline of useful attention, reverts to the state of a child—the sensual and unconscious instinct reassuming its sway over the voluntary and intellectual instinct ; the man returns gradually to the infant state of looking for the mere pleasure of seeing, of listening for the sake of hearing, of feeling for the sake of gratifying the touch, of moving and walking for the sake of the more or less agreeable sensations which these automatic actions produce. How many people are there not whose days are passed in vacuity, that is to say in nothing higher than the unconscious functional actions of the instincts of sense, which have gradually become transformed into barren habits ! That which is a constant necessity for a little child, who after slight efforts at attention and intellectual operations requires to rest in less engrossing sensations, is only necessary to the adult at intervals, in order to recover from nervous or muscular fatigue, or in case of illness.

*The Instinct of Nutrition.*—The appetite for nutrition, or the instinct of nourishing oneself, which occupies the first place among special appetites, is not one of those tendencies whose stimulus is always present in the visceral organs, as the need of breathing, for instance, is from the moment of birth ; this stimulus has to be supplied from

•

without, and the state of the viscera to which the appetite for nourishment corresponds is reproduced at regular intervals. Even in the later stages of uterine life, the intimate union between sensory activity and visceral wants is established relatively to this instinct; it is because its functions are too delicate, too important, and too complex for the organs destined to perform them to be able to come into play at the moment of birth—a time liable to a multitude of accidental variations. Hence the following facts, related by Houzeau, are by no means improbable.

"Mammals and birds absorb nourishment before escaping from their integuments; the chicken consumes the white of the egg, the child sucks up the water of the amnios, as also do calves and the greater number of mammals. This fœtal alimentation is the cause of the often very copious motions of new-born animals. But calves go still further. It is an ascertained fact that they lick themselves before the time of birth, pieces of their own hair having been found in their stomach with the water of the amnios. . . . The instinct of nourishment is thus as innate as the need of taking nourishment."[1] We should even add, that this instinct is already specialized, "that it does not impel the infant indifferently towards all objects alike, but towards a certain class of objects." After the first months of milk diet, children show very marked omnivorous tendencies, with characteristic differences of likes and dislikes according to the individuals.

The instinct of nutrition is innate; and so, up to a certain point, is the faculty of nourishing oneself, but it is not perfect on the day of birth. Some animals, as Bastian says, exhibit this faculty "almost immediately after birth, and without making any previous abortive efforts."[2] The wild boar and the chicken, for instance. But in the case of animals who do not attain, while yet in the oviduct or the uterus, the necessary development for the exercise of certain faculties, the latter do not appear, or rather are

[1] *Etude sur les Facultés Mentales des Animaux*, t. i., p. 193.
[2] *The Brain as an Organ of Mind.*

not developed, until a few days or weeks after birth. A child manages the operation of suction, though awkwardly it is true, a few hours after birth ; but it cannot masticate or take hold of other food adapted to its species any more than a new-born kitten can catch or devour prey. And just as a kitten does not learn to lie in wait for prey, or to seize its food and masticate it until certain parts of its organism have become sufficiently developed, and birds do not attempt to fly away and seek food for themselves until the right organs are ready, so children only learn to take hold of and eat an apple, a piece of bread, a cake, etc., when their organs of mastication have reached a certain stage of maturity. I may add, that the forces which suffice to produce the stimulus of instinctive action in infants, as in kittens, and even in young birds, result at first merely in more or less successful attempts. The power of feeding oneself, like that of walking or flying, requires always a longer or shorter education, in which the initiative of the young being needs to be helped by the example and encouragement of grown-up people.

*The Instinct of Sleep.*—The appetite of sleep manifests itself in quite a special way, as a negative tendency, if I may so express myself, as a need for the cessation of activity. Hence the movements and actions which produce tranquillity and sleep are monotonous and slow ; the natural rocking of branches for birds, the rhythmic movement of the head in horses when standing upright, and the rocking of the arms or cradle, accompanied by a monotonous song, for children's sleep, is interesting to study from the double point of view of physiology and psychology.

The physiological cause of sleep is unknown. On whatever hypothesis it be explained, whether cerebral congestion or cerebral anemia, we can at any rate easily observe the almost constant characteristics of this phenomenon. It takes possession, gradually in adults, and often abruptly in children, of all the different organs : first of all the muscles of the limbs, the arms and legs, become fixed in the position they happen to have assumed ; after the limbs, the voluntary muscles of the trunk relax themselves into a state of more or less complete flexion.   It has also been noticed that during sleep respiration and pulsation become rarer.   But what is the mental condition of the animal while asleep?   While the **nutritive functions** and reflex movements are still going on, does the brain cease to operate?   Does sleep only suspend a portion of the phenomena of psychic activity?   We know nothing about it.   It is probable that the abolition of mental states is never absolute, that it never reaches all the regions of the brain, even when sleep is profound.

It is at any rate certain,—both memory and the nature of dreams prove this,—that the cerebral hemispheres have a great tendency during sleep to recommence their functions, though always in an incomplete manner.   We know also that the impressions which come from the viscera, or a very slight hindrance in the circulation or respiration, or too strong a muscular pressure, or repletion or vacuity of the intestines, or even perceptions from the external world coming abruptly and disagreeably, will give rise to very painful ideas and emotions, and cause corresponding screams and movements.   Although dreams do not appear to be excluded, even from profound sleep, they are generally fatiguing to the organs.   Sleep means the reparation and restoration of the forces of the body, and the incoherent activity of the dream state is a premature expenditure of the force intended for the activity of the morrow.

But when dreams are light, intermittent, and pleasant, when they do not hinder the complete repose of the principal organs, and when they only exercise the cerebral organ moderately and without tiring it, they are entirely negative from a physiological point of view, while at the same time they are highly favourable to the intellectual and moral development.   In the absence of any actual perceptions, past perceptions work themselves out under the law of association, and free from the control and the obstacles of reality.   The mind, as it were, isolates itself from the external world in order to abandon itself freely to its work of ideal incubation and digestion.

It is then, and perhaps better then than in our waking hours, that the fortuitous association and dissociation of images produce those abstractions by which individual forms are detached from masses, details from the whole, and qualities from objects ; there is no doubt that in dreams a child's brain is crossed by images as vivid in themselves as they are unlike absolute realities : hallucinations of sound, colour, and tactile impressions, of muscular sensations, of forms isolated or grouped together, of objects that resemble each other, and objects that are differentiated, and still more series of actions accomplished by a greater or less number of actors, and giving rise to reasonings and sympathetic sentiments. All the intellectual operations, all the emotional faculties, are exercised in dreams with all the more ease and utility that they act solely on their own resources ; in this state the mind works on its own acquirements and ideas, we may almost say, on itself. Thus it may be said that sleep is for children general repose of muscular, sensorial, and cerebral activity, though with intermittent returns of activity, all the more agreeable and serviceable as the dreams are rare and light. We know, moreover, that a child of two or three years old dreams more than one of six months or a year, a child of from six to ten years, more than one of three, the adult less than the youth, and the old man less than the adult. The frequency and the vivacity of dreams, pre-supposing of course the normal condition of the subject, appears to be in proportion to the psychic excitability or activity.

Dreams are not only an important fact of psychic life, interesting to our intellectual faculties, their influence extends to our sentiments and even to our morality. The mental states which are produced, with or without consciousness, during sleep, are the consequence of and the preparation for certain states of our waking hours. " It is possible," says M. Ch. Lévêque, " that the cheerful or sad humours of the day are a faint repetition of the agitations experienced in sleep, and that all the workings of the mind during the night may help to produce certain actions of the day." There is indeed no doubt that remembrance is not the only trace dreams are able to leave after them. According

to the nature of the dreams which it has had during the
night, the child is more or less cheerful during the day,
more or less inclined to be good and obedient.

I have said enough about the instinct of locomotion in
the chapter on movements, to make it unnecessary to go
over it all again here. The instinct of utterance I shall
treat of in the chapter dedicated to language. The instinct
of sexuality will be most fitly considered in a chapter devoted to modesty or to the moral sense ; though it will not
be amiss to devote some attention to it in this chapter concerning special tendencies.

*The Sexual Instinct.*—The localization of the sexual
instinct in the brain is far from being determined. The
hypothesis of Gall, who asserted that a constant relationship
exists between the development of the cerebellum and the
sexual appetite, has been entirely confuted by the facts
elicited by clinical observation and human pathology. According to Ferrier, the organic needs which constitute the
basis of sexual appetite, centre round a special form of tactile
sensation which may be supposed to have its centre in close
relation to the hippocampal region. But it is only conjec-
ture ; he affirms nothing. He adds, moreover, resting his
supposition on the power which some odours have to excite
the sexual instinct, that a region in close relation to the
centres of the sense of smell and to the tactile sensations
might be considered as the probable seat of the sensations
which constitute the basis of sexual appetite.[1] The centres of
this sensation, according to the same writer, " are probably
placed in the regions which unite the occipital lobes to the
infero-internal region of the temporo-sphenoidal lobe. As
the reproductive organs in women form such a preponderant
element in their bodily constitution, they must correspond-
ingly be more largely represented in the cerebral hemispheres,
a fact which is in accordance with the greater emotional
excitability of women, and the relatively larger development
of the posterior lobes of the brain." [2]   Thus we are reduced

---

[1] David Ferrier, *The Functions of the Brain*, p. 198.
[2] Ibid., p. 263.

to hypotheses with regard to the localization of this instinct, and it is not necessary to inquire whether the brain of a little child possesses, in any degree of development, the cerebral organs of sexual appetition. But we must not overlook the fact that this instinct is not exclusively characterized by the reproductive tendency. Its chief characteristic is the appetition of the sensations whose unconscious object, or rather, perhaps, whose result, is the multiplication and preservation of the human race. I quote here the words of a recognised authority on comparative psychology.

"Modern anatomists have proved that the disparity of the sexes is much less radical than we are inclined to think. The type is the same for all the individuals of a species. The male mammifer, like the female, has breasts, which only lack development. The analogy is continued, if not between the relative proportions of the different parts, at least in the general plan of structure, down to the genital organs. Everard Home goes as far as to think that at the beginning the germ may be indiscriminately endowed with either sex, and that its being male or female would depend on ulterior circumstances of a simple nature, such as accidents of impregnation. . . Even in superior animals, the sex, though it cannot be changed, is not such an exclusive thing as is generally thought. At birth, and during all the first period of life, the sex can only be distinguished by the structure of the genital organs. It is only later in life that other characteristics appear, the beard, the breasts, etc. The male bird decks himself out in the most resplendent plumage; and those mental affections which are allied with sexual phenomena, and which have hitherto lain dormant, or almost so, now burst forth in all their strength." [1]

The same author cites a number of examples taken from the mammifer species, and tending to show that the deviation of the sexual instinct,—going as far as to anomalous confusion of sex and age, and the exclusive pursuit of sensations connected with this instinct,—is not special to human

---

[1] Houzeau, *Etudes sur les Facultés Mentales des Animaux*, tome. i., pp. 274, etc.

beings.  In recording these physiological observations, the
great naturalist has no other object than to explain the
brutality of the passion of love in different animals.  They
enable us to understand the like excesses in man, who is
not necessarily a reasonable animal, but an animal who can
and ought to be reasonable.  Without dwelling on these
delicate questions, which frequent criminal scandals and the
success of a certain class of novels force on the consideration of moralists and legislators, I shall content myself with
pointing out to parents and educators who are ignorant of
them, the possible dangers resulting from the deviation and
corruption of the sexual instinct, even as regards quite young
children.   Let me again quote a few passages from a book
of high moral purport, which grown-up people may study
with profit both to themselves and to children of any age
who may come under their care.  The quotation refers only
to the moral interests of young children.

"The following points, bearing on the moral education of
childhood and youth, must be considered by all parents
who are convinced of the saving value of sexual morality,
viz., observation of the child during infancy, acquirement
of the child's confidence, selection of young companions,
care in the choice of a school and of studies which will not
injure the mind, the formation of tastes, outdoor exercise,
companionship of brothers and sisters, the choice of a
physician, social intercourse and amusements.  These various points require careful consideration.

"The earliest duty of the parent, is to watch over the
infant child.  Few parents are aware how very early evil
habits may be formed, nor how injurious the influence of
the nurse " [why of the nurse only ?] " often is to the child.
The mother's eye, full of tenderness and respect, must
always watch over her children.  .   .   . This watchfulness
over the young child, by day and night, is the first duty to
be universally inculcated.  Two things are necessary in
order to fulfil it ; viz., a clear knowledge of the evils to
which the child may be exposed, and tact to interpret the
faintest indication of danger, and to guard from it without

allowing the child to be aware of the danger." [1]   Dr. Elizabeth Blackwell adds, in a note in the Appendix, the following melancholy considerations :—" Terrible instances of this may be seen in Trélàt's medical work, 'La Folie Lucide;' and Lallemand and other French surgeons report numerous cases of fatal injury, done even to nursing infants, by the wicked actions of unprincipled nurses.   I have myself traced the ill-health of children in wealthy families to the habits practised by confidential nurses, apparently quiet respectable women !   Abundant medical testimony confirms these observations." [2]

---

[1] *Advice to Parents on the Moral Education of their Children*, p. 85 (Dr. Elizabeth Blackwell).
[2] Id., Ibid.   Appendix.

# CHAPTER V.

I CANNOT resign myself to seeing in a little child, even when only a few days old, a mere machine or automaton. A learned man to whom I had expounded, without convincing him, my ideas on the direction that should be given to the faculties of young children, sent me by way of refutation the following description of his infant son, then about two months old :—" He is a thorough little animal, voracious to excess, and never quiet except when asleep or at the breast. I could never have believed that a little child was so absolutely an animal, with no other instincts than that of gluttony. To avoid being completely disgusted, one has to remind oneself that in a few months there will be some gleams of intelligence, and that the creature will begin to show some likeness to a human being." All at once, I suppose ! By one knows not what miracle of nature ! My friend received from me a variety of further observations which, however, had no greater success. He wrote again : " I believe still in the pure animalism of the infant, and to give a real idea of the voracity of this age, I can only compare it to a larva, always eating without pause or rest." Setting aside for the moment some details which appear to me to be doubly calumnious, both towards children and animals, I must allow that my friend has well observed and well described the state of a young child subject to the tyrannical need of food. M. Luys has observed the same state of things ; but he, like myself, has also observed something further, and this something is of the greatest importance : " His organic appetites are gratified by the milk he sucks, and he feeds himself organically, like an organic cell, which borrows from the surrounding medium the materials which suit it. But at the same time he expresses the satisfaction he feels in his own manner; he smiles on seeing the breast which yields him his nourishment and life, and from that time his natural sensibility is thrown into agitation, his *sensorium* is affected. He rejoices because he remembers, because he has retained a memory of the satisfaction of his physical appetites." [1]

*Sentiments connected with Taste.*—A child's most vivid sentiments are for a long time after its birth those connected with the sense of taste. The need of nourishment dominates for a long time over all the others, even that of movement; it manifests itself the first and is the most persistent. The emotions connected with this incessant and imperious want are the most agreeable ones that children experience. It is through the satisfaction of their appetite that they learn to know and to love first of all the breast they suck, or their feeding-bottle, and secondly the hands, the face, the voice, the eyes, the smiles, the caresses, the entire person of their nurse or mother. Their first affections are those of an epicure; their first feelings of gratitude are awakened by the stomach; they test their first tactile experiences, as much as possible, by the sense of taste. If one puts an object of any kind into the hands of a child of six months, he will touch and feel it for a few seconds without seeming to learn much from the tactile impression, and then, if he is strong enough, he will carry it to his mouth, and experiment upon it with the organs of taste. His nurse's finger, a bit of rag, a stick, a box, fruit, flowers, etc., anything and everything, great or small, pleasant or unpleasant, all goes up to the mouth. *Pretty to look at*, and *good to eat*, are synonymous terms to babies; a pretty picture, the colours of which first attracted it, is seized hold of, and, like everything else, put into the mouth. Even after they have learnt by frequent experiments that all objects have not a pleasant taste, they

---

[1] Luys, *Le Cerveau et ses Fonctions*, Eng. Trans , p. 127.

only mistrust, at first sight, those which are notoriously offensive to them. Later on, after the age of four months, the desire to stop the itching of the gums becomes, no doubt, an additional reason for the constant movement of the hands to the mouth ; but the chief cause of it is the excessive excitability of the functions of taste, and of the ideas and sentiments connected with this organ.

The sensations of joy and pain connected with taste continue to be the dominant ones during the first months ; but they are not the only ones that children feel.

*Fear.*—The automatic instinct of fear exhibits itself in infants from the very first. We may even, moreover, see obscure manifestations of this instinct in the tremblings produced in the fœtus by any sudden terror in the mother. A lady who had had a great shock three months before the birth of her child, felt the child move convulsively within her. The baby only lived three months ; and during this time it used frequently to give those violent starts, without any external cause, which characterize excessive fear. These are indisputably the effects of congenital imaginativeness, against which a mother and those around her should guard as much as possible.

As for the tremblings, the screams, the cessation or precipitation of breathing, which are common symptoms of fear in new-born children, their cause is often so slight that it is not always possible to foresee or prevent them. Sudden sounds and sights of all sorts clearly distinguished, disturb a child's rest. At three months, and even earlier, the mere sight of a strange face will sometimes so agitate a child and affect its breathing, as to make it seem on the point of suffocation. Darwin has noticed signs of fear in infants during the first weeks, at the slightest unexpected noise, and later on at any strange noise or attitude. He speaks also of the fear felt by some children at a more advanced age, on finding themselves in the dark ; but he does not tell us whether he considers this tendency hereditary. We think, however, that such is his opinion, for he attributes (what to my mind seems exaggerated) the fear felt by his child in the Zoological Gardens at the sight of large animals " to the hereditary effects of real dangers and abject superstitions which prevailed at the period of savage life." With all deference for the opinions of this illustrious physiologist, I cannot help asking whether a little baby's fright at the sight of an enormous animal may not be partly due to its own sense of comparison ; and whether it is necessary to go back to the experience of its savage ancestors to explain this particular manifestation of a tendency which is hereditary only in so far as it is general.

The following observation of Charles Bell may no doubt be applied to the age of six or seven months :—

" If an infant be laid upon the arms and dandled up and down, its body and limbs will be at rest as it is raised, but in descending it will struggle and make efforts. Here is the indication of a sense, an innate feeling, of danger ; and we may perceive its influence when the child first attempts to stand or run. When set upon its feet, the nurse's arms forming a hoop around it, without touching it, the child slowly learns to balance itself and stand ; but under a considerable apprehension ; it will only try to stand at such a distance from the nurse's knee, that if it should fall, it can throw itself for protection into her lap. In these its first attempts to use its muscular frame, it is directed by a fear which cannot as yet be attributed to experience."[1] If a child tumbles down while trying to walk, it sometimes gives up the attempt for a long time. But there is as great difference in this respect in individuals of the human race, as in the young of animals.

Like little children in their first efforts at difficult games, or in their first gymnastic exercises, some puppies exhibit remarkable boldness, which no amount of tumbles can overcome, and others a laughable degree of cowardice and prudence.

Fright is less often caused in children between the ages of three and ten months by visual than by auditory impressions ; with kittens, after the fifteenth day, the reverse is the case. A child of three-and-a-half months, in the midst of

---

[1] *The Hand*, p. 234.

the alarm of a house on fire, and surrounded by flames and tottering walls, showed neither astonishment nor fear; he even smiled at the woman who was taking care of him and keeping watch over the furniture while waiting for his parents. But the sound of the bugle, and of the firemen coming up, and the noise of the engine wheels made him tremble and cry. I have never seen a child at this age startled by lightning, however vivid; but I have seen many terrified by the sound of thunder. On this point my observations are in opposition to those of Rousseau and Herbert Spencer. The former thinks erroneously, with Locke, that fear is a sentiment derived from experience of hurtful or dangerous things, and that, in the absence of very startling impressions, it does not become developed in children. " I have noticed," he says, " that children are rarely frightened at thunder, at least if the claps are not too violent, and do not hurt the ear; otherwise this fear does not disturb them until they have learnt to associate thunder with danger and possible death."[1]   Herbert Spencer's opinion is as follows:—

"It happens, no doubt, that a child may be seized with terror at a clap of thunder; and an ignorant person will regard a comet with superstitious terror. But claps of thunder and comets are not every-day phenomena, and do not form part of the usual order of things."[2]   The fact is, that certain children, in their first months, are frightened by certain very sharp or very sonorous sounds, and above all, by unusual sounds. It is also true, as Locke and Rousseau have observed, that the more a child becomes accustomed to any sounds, the less it will be frightened at them. A child who was very much frightened by thunder at the age of six months—say in May—would not be so frightened five months later, at the end of September, having heard the same sound several times, and having become familiar with it. But at a still later age, owing possibly to bad training, this fear will reassert its dominion.

There is a kind of natural fear, organic and hereditary,

---

[1] *Emile*, livre I.          [2] *Principles of Sociology*, vol. i.

the result of anterior experiences, and which is a safeguard to the young infant against certain very real dangers, of which it has not yet had any experience. This is the reason why fear is stronger and more easily excited at five years old than at three, and at three years than at six months, and also why it is more apt to be aroused in little children by auditory impressions than by visual ones. The anterior life of civilized man has rather predisposed the race to listen for dangers which are near at hand, than to be on the look-out for distant ones; *i.e.*, the ear has been more trained to keep watch than the eye. And accordingly, setting aside all individual susceptibilities, which are the fruits of special heredity, we find that in the inexperienced infant fear is excited rather through the ear than the eye. It is natural that the reverse should be the case in animals, so organized as to perceive danger from afar. Thus, though I have never come across a child who was frightened at the first sight of · fire, I have found the contrary to be the case in several kinds of domestic animals—dogs, cats, chickens, and birds for instance. A chicken found lying half-dead in the garden was brought indoors and placed near the fire; and, in spite of its feeble condition, it quickly hopped away from it. It was brought back again and placed on a stool, and after a good deal of coaxing and petting it began to lose all fear, shut its eyes, and fell asleep. It was then left to itself; and whether that it was too feeble to fly away, or that it had become happy in its new situation, it remained perfectly still there all the rest of the day and the following night. The next day it had quite recovered, but it came back of its own accord to take up its position on the footstool. For several years past I have given a home to a stray cat. She was about a year old when I first took her in. A few months after, when the cold weather set in, I lighted a fire in my study, which is also the cat's sitting-room. At first puss looked at the flames with a frightened expression. I made her come near the fire, but she bounded away and hid her-self. I had a fire in the room every day, but it was not until towards the end of the winter that I could induce the cat to remain on a chair near it. By the following winter, however,

F

her fears had disappeared, and she would as a matter of course seat herself on a chair in front of the fire. Noise of any description at once puts her on her guard—with her eyes, not with her ears; the sound of thunder or of a heavy waggon causes her to look up suddenly at the ceiling, and then to listen at the window; the reason of this is, that the shadows of objects passing along the street in the evening often flit across the ceiling of my room, as I only have my shutters half-closed. The conclusion, then, that we arrive at is, that there are predispositions to fear which are independent of all experience, but which the gradual accumulation of experience lessens very considerably, and that in the case of children they are chiefly connected with the sense of hearing.

*Anger.*—During the first weeks of existence children's instinctive mode of expressing the pain which any object inflicts on them seems to consist only in screams and movements of resistance. But when about two months old, they begin to push away objects that they do not like, and have real fits of passion, frowning, growing red in the face, trembling all over, and sometimes shedding tears. At three months old, they begin to experience the feeling of jealousy, which is shown in tears, screams, and contortions, if one pretends to be going to take away their feeding-bottle or any other object of their affections. They also become very much irritated if they cannot at once get at their mother's breast, or when being washed, or having their clothes changed, or if their wishes are not guessed and satisfied quickly enough. At six months they will scream with impatience if their toys are taken from them. This may be either owing to an inborn instinct of proprietorship, or because of the amusement which their toys afford them.

At this same period their movements and cries during sleep appear to indicate painful dreams. Towards the age of one year their anger sometimes exhibits itself in hurtful actions in which we see, so to speak, the germ of the passion of revenge. They will beat people, animals, and inanimate objects if they are angry with them; they will throw their toys, their food, their plate, anything, in short, that is at hand, at the people who have displeased them, or simply at the first person near, when it is the objects that have caused displeasure. Thus anger has its origin, and that at a very early age, both in simple and complex sentiments, and is expressed either by simple and automatic actions or by complex ones acquired personally.

If the theory of evolution is true, it is necessary for the young human being to pass through, in a certain gradation, all the principal stages which have brought his ancestors from animalism up to the first beginnings of civilization. It is but natural that a child should at one moment reproduce this ancestor, at another resemble that savage, with whom many would identify primitive man. Now, irascibility is one of the special characteristics of inferior races. Irritability and impulsiveness are, with few exceptions, fundamental traits of all these races.

"Spite of their usually unimpassioned behaviour, the Dakotahs rise into frightful states of bloody fury when killing buffaloes ; and among the phlegmatic Creeks, there are "very frequent suicides, caused by trifling disappointments." . . . Passing from North America to Asia, we come to the Kamtschadales ; of whom we read that they are "excitable," not to say (for men) hysterical. A light matter set them mad, or made them commit suicide." . . . Among the Negrittos the Papuan is "impetuous, excitable, noisy ;" the Fijians have "emotions easily roused but transient." . . . The Tasmanians "quickly change from smiles to tears." . . . The Fuegians "have hasty tempers, and are loud and furious talkers." "There are the Australians, whose impulsiveness Sturt implies by saying that the 'angry Australian *jin* exceeds the European scold,' and that a man 'remarkable for haughtiness and reserve, sobbed long when his nephew was taken from him.' "[1]

This impulsiveness and irascibility are in our children legacies inherited from our primitive ancestors : they are, as it were, instinctive weapons of defence, and native instruments

---

[1] Herbert Spencer, *Principles of Sociology*, pp. **63** etc.

of self-preservation. It appears then *à priori* to be established, that systematic evolution, or education, should preserve
this force while disciplining it. But experience alone can
teach us what may be the consequence, either in infancy or in
the future, of combating or encouraging these sentiments,
and consequently what place we should allow this moral
factor in education.

Anger is legitimate in young children when it expresses
unconscious revolt against the first sufferings of life—convulsions, colic, pains of teething, discomfort produced by fever,
or by the want of air, of locomotion, or of sleep. The
screams and movements in which they vent their anger dis-
tract and relieve them, in a certain measure, from the feeling
of pain ; and to the parents or guardians of the child they
are warnings dictated by nature herself. In like manner,
when a child, in its first awkward attempts at speech, has
given a wrong idea of its meaning, it is quite justified in
screaming, beating the ground with its foot, and showing
indignation at being so badly understood when it thought
it had spoken so well. It has still more right to be angry,
when, after having been accustomed by its nurse to any bad
habit, such as being rocked to sleep in its cradle or in any
one's arms, or being put to bed with a light in the room,
these habits are suddenly discontinued ; still more so again,
when, without regard for its delicate sensitiveness, we try to
force it to do something which, either from nature or habit, is
repulsive to it, such as swallowing a bitter draught, or undergoing some punishment or privation without complaining,
or kissing a person it dislikes. Another case, mentioned by
Rousseau, in which a child's anger is perfectly legitimate, is
when the nurse beats it for crying, and, an additional form
of suffering being thus added to that which he probably
experienced before, he expresses both by loud screams and
rage.

There are also certain forms of impatience which betoken
a frank and generous character, and which are closely allied
with the budding of the earliest moral virtues. A young
child, for instance, who delights in walking alone, although
at the risk of falling, will get extremely angry if any one

should persist in trying to help him. At a later age, again, suppose a child to repeat in his mother's presence some foolish joke which the servants had always been very much amused at, but in which his mother sees nothing to laugh at, his face will get very red and he will close his lips firmly, evidently feeling that he has exposed himself to ridicule and appearing quite irritated with himself. Again, at about two years old, when a child is suddenly punished severely by some person who had before been generally indulgent to him, he will fly into a passion for a punishment which he would have received submissively had it been inflicted by his parents. Once more, if he sees two children fighting in the street, he will run up to them with clenched fists and crimson cheeks, and try to separate them. Are not all these ebullitions of temper invaluable guides to the knowledge of a child's character and helps in its moral education?

But if anger has its good sides, its use and justification, it has also its evils and abuses. It is often the outcome of caprice, jealousy, hatred—of all the hostile passions, combativeness, destructiveness, and vengeance. It is the two-edged sword of human wickedness, which wounds the striker as well as the victim. If indulged in too frequently, it will injure the moral and physical development of the child, who ought always to be surrounded by an atmosphere of peaceful serenity, and in whom we should endeavour to maintain calmness and tranquillity of spirit. Outbursts of anger may have specially disastrous effects on children predisposed to convulsive maladies at an age when the muscular system is not sufficiently developed to counteract shocks to the nervous system. What is more likely to hinder the growth of good-humour or docility in a child, than the habit of getting irritated at the slightest cause—because an object he tries to take hold of slips out of his hands, or because he is given something to eat or to play with that he does not like, or because a stranger speaks to him or kisses him? What less pleasing spectacle can there be, than that of a pretty little child a year or two old, who habitually vents its anger on the furniture, the books, flowers, fruit, or food, the cat or the dog, its nurses, or even its parents?

I have seen a capricious little creature eleven months
old put herself in a violent temper because she could not
succeed in seizing hold of her grandfather's nose ! Another
child I know of had a beautiful doll, of which she was very
fond; her parents took her once with them to Cauterets,
and on getting out of the carriage she saw another child
with a doll just like her own ; instantly there were screams
and paroxysms of rage ! She flew upon the child, scratched
her, beat her, and bit her ; and she had to be carried away
by force. Her fury was so great that she was quite ill from
it for several days. Another little girl of the same age had
such fits of passion every evening when her mother was
putting her to bed, that the neighbours would sometimes
come in and help to quiet her. There was one person
whom she specially dreaded, and the sight of whom was
sufficient to quiet her : this was a gentleman with a loud
voice and a long beard, who sometimes whipped her in her
cot. A little boy fifteen months old used to bite his mother
when she put him to bed. Another child, three years old,
who had been sent away from the dining-room on account
of his naughty behaviour, came back soon after and laid
himself down on the floor across the doorway, throwing out
his arms and legs, and screaming at the top of his voice.

Having thus pointed out the use and the abuse of anger, it
seems to me that we may, on *à posteriori* as well as on *à
priori* grounds, place this passion among that class of animal
sentiments which it would be waste of time to endeavour to
exterminate, but which need to be carefully directed and controlled. We should recognise in the passionate temper one
of the most fruitful principles of human activity, one which,
if united with sympathy, will lead to acts of self-devotion
and may help in the formation of moral habits by obliging
the child to examine himself and his own actions, or by
inspiring him, as far as his tender age permits, with a germ
of—

> . . . ces haines vigoureuses
> Que doit donner le vice aux âmes généreuses.

*Jealousy.*—The instinct of jealousy, common to all animals, but unequally distributed amongst individuals of the same species, manifests itself in very different ways and circumstances. It is not always the sign of very acute sensibility or of strong personality, for it shows itself very markedly in young children and in adults of a calm and equable temperament. Sometimes it will burst into flame, like latent fire. Sometimes it smoulders on like burning ashes. Love and affection are its most violent excitants ; but any trifling cause may also give rise to it. A cat or a dog will be jealous of each other about their food, or about a favourite place, a plaything, or a caress. A sparrow tamed by a lady was jealous of the cats when its mistress fondled them, and of the visitors who came to see her; its attitudes and cries plainly betokened this. In like manner a child will show jealousy if any one approaches its nurse, or touches its bottle, or is caressed by its mother. A child of fifteen months was evidently jealous if sugar or dessert was given to its nurse.

This feeling is roused by very different objects, and is sometimes confounded with envy and the desire of appropriation and imitation. Children often want things not so much for the sake of having and enjoying them, as because they do not like to see them in the possession of others. And what applies to things applies also to persons. A child of fifteen months used to enact very curious little scenes out of jealousy. If his father and mother kissed each other in his presence, he would run up and try to separate them, scolding and pushing away his father, who was by no means the favourite. The same child, at this age, could never see anything in anybody else's hand without asking for it, or trying to touch or take hold of it ; nothing could ever be done in his presence, without his wanting to meddle in it. In the kitchen he must have a knife, or something like a knife, and pretend to be at work with the parings of the vegetables while the servant is preparing the dinner. When his older brother is writing, he insists on having a high chair at the table and some paper and a pen, and then he fancies that he is doing the same as his brother ; once he gravely asked for his father's razor, that he too might shave. Thus

we see that the proprietary and imitative tendencies enter
largely into the displays of envy and jealousy.

Fathers, by the way, need not be troubled by the preference their little children generally show for their mother or
nurse. This is quite natural, and more favourable than
otherwise to their moral and intellectual development.
Mothers, on their part, need not be unduly distressed at the
inconstancy of their little hearts. They will right themselves
in time, provided the parents do their duty by their children ; and manifestations of jealousy in either father or
mother would be a very bad example to a child.

There is always, no doubt, a little self-love, if not vanity,
mixed up with the jealousy a child feels towards its brothers
and sisters, especially with regard to the marks of tenderness and attention which are shown them. Tiedemann
says of his son, at the age of twenty-two months, "Jealousy
and vanity became stronger and stronger; if any one praised
his little sister, he instantly came up to be praised also ; he
always endeavoured to get away from her whatever had been
given her, and he would even hit her on the sly." When
this little sister was born, the boy had shown signs of displeasure ; he wanted to beat her whenever he saw her on
his mother's lap or in his own cot, because it was disagreeable to him to see anything which he had possessed
exclusively for some time taken away from him.

One of my nephews, at the age of three years, used continually to talk of the little brother he was soon to have.
" I shall love him so much," he would say at every instant.
But when he saw the baby taking up his mother's lap and
kisses and caresses, and his father's care and attention, he
expressed his annoyance loudly. He even said to his
mother one day : "Won't little Ferdinand soon die?"
When the baby began to walk and talk, the elder child would
torment him in hundreds of naughty ways, beating him,
dragging him out of his chair in order to take his place,
shouting in his ears, calling him naughty and ugly, taking
away his toys, and mimicking his way of talking and walking.

*Sentiments of various kinds.*—When experience has taught

a young child to know a certain number of objects as having
the power to afford him pleasant or painful sensations of
sight or touch, his waking hours begin to be more and more
divided between his meals and his playthings. To the instinct of hunger or greediness, still dominant but no longer
exclusive, there are now added fresh wants, which the Scotch
call intellectual, because they imply a certain growth of the
intelligence.

Here is a child of eight months, of ordinary intelligence.
He is interested in a number of objects which have nothing
to do with his palate, and which he only carries to his
mouth accidentally. These objects are instruments of play
and study for the child. He handles them, turns them
round and round, knocks them down, sets them up again,
throws them away, fetches them back, crawls after them on
all fours when they are out of his reach, knocks them one
against the other, puts them inside each other, thrusts his
hands into them, piles them up in heaps and then knocks
them down; in short, disports himself with them and learns
from them in a thousand different ways. Sight and touch,
which before seemed generally mere auxiliaries of taste,
now act on their own account; the original synthetic condition of the functions has now given place to an analytic
condition which gains daily in strength and delicacy; the
concentric circles of sensations, perceptions, judgments,
sentiments, go on expanding; and henceforth my friend the
savant is able to admire the *little animal*, who rises day by
day and hour by hour to the level of a *little man*.

Take another child of eleven months. He is passionately
fond of his bottle, which is for him the embodiment of exquisite enjoyment; but he has made acquaintance with a
certain number of other eatables—soup, bread, meat, cakes,
fruit, etc.—which he seems to like as much as his first food.
However, when his appetite is satisfied, and he is taken
back to his toys, one sees that he enjoys them just as much,
if not in the same manner, as he did eating and drinking.
He shows the same desire to seize hold of them, the same
attraction towards them, the same distress if they are taken
away; his expressions, gestures, and attitudes of delight prove

that they afford him equally agreeable sensations. Then again I perceive that he shows love for his mother, his nurse, his sisters, his aunt—in fact, for all the people who feed, pet, talk to, and amuse him. Moreover, there are evidently different degrees in his affection. He seems more pleased to go to the arms of his mother or nurse than to his little sister, who hugs him so awkwardly that she sometimes makes him cry, or to his aunt, who means to be very loving and caressing, but who does not look so, and who, in addition to a repelling countenance, has a shrill voice. This child brightens up at the sight of a young or pretty face, but shows very little interest in old or ugly ones, or faces covered with veils. His affections vary according to the nature of objects and the sensations they afford him, and according to the character, manners, and actions of different persons ; he has also his dislikes, both for persons and for things. He gets exasperated when a little neighbour, seven years old, who has played him several tricks and made faces at him, comes up to him to kiss him. One of his uncles often brings with him a little black dog, much given to barking ; and the mere sight of this animal distresses the child. Soap and water, and towels, the rod, and the enema syringe, he looks upon as personal enemies. Inanimate objects have a large share in the sentiments of little children. The pleasure and the pain which these objects cause to some of his senses are the germ of all these affections and repulsions. But curiosity and the incessant need of new and vivid emotions are his stimulants, and furnish daily food for his affective sensibility.

The affections of a child, however, like the curiosity which excites them, are very transitory. They glide from object to object, from person to person. Owing to their ignorance and feeble power of attention, they are unable to occupy themselves long with one person or object which cannot vary and be metamorphosed every instant to suit their restless curiosity. Here is a child of ten months. He confuses flowers with other inanimate objects; a piece of rose-coloured paper excites in him as vivid and durable a feeling of curiosity and pleasure as would a beautiful rose. When I sniff a rose in his presence, and invite him to

imitate me, **saying to** him, " It smells so nice," he draws in
his breath and looks pleased, showing that he is sensible
of the **sweet** smell of the rose ; but the sensations which
perfumes excite in us are so quickly effaced that the artificial
habits of adults are necessary in order to appreciate them,
and to like flowers for the sake of their smells.   These sub·
tle charms, so keenly enjoyed by grown-up people, soon
cease for children.   With regard to these fragile works of
**nature, their pleasure is** still more fragile, especially when
they hold them in their hands.   They will jump with delight at **first** seeing and smelling them, but will very **soon**
leave them for something else.   The infant organization,
more feeble than our own, is subject, almost without exception, to the law **of nature** according to which the most vivid
sensations are **the least enduring.**   This is the **reason** why
objects which excite in children sensations keenly painful
or pleasant, never please or distress them for any length of
time.

*Animal Sympathy.*—Animate **objects have rather more**
power than inanimate ones of arresting the attention **and**
awakening curiosity, and hence of exciting the emotions of
children.   Dogs, cats, sheep, birds, chickens — all these
creatures are *par excellence* their objects of recreation, instruction, and affection.   And it is not strange that it should
be so.   They afford all the gratifications of sight and hearing combined, all the various pleasures of touch, and that
latent voluptuousness which follows the satisfaction of the **appetite of movement.**   Added, also, to the perpetual **renewal**
of curiosity excited by these animals, are the no less **powerful**
influences of animal sympathy.   How intense is a child's
delight when almost strangling one of these good-natured
creatures in its eager grasp.   It seems like its **own** life, its
own personality, vibrating in those organs which beat with
the same movements as his own.   This feeling, which is
hereditary **rather** than acquired by sympathy, sometimes has
phases of superior excitation : **the** games, the caresses, the
screams of delight, the gentle purrings, the wailings and
moans of a cat or a dog, all this combination of sympathetic
sensations stimulate the curiosity and excite to the highest

pitch the affections produced by the increased gratification
of the senses of sight, hearing, touch, and movement which
the presence of these animals affords children.

This sympathy for animals, however, does not include
moral sympathy.   The child of a neighbour of mine, who
is two months old, and who plays all day long with the cat
and the dog, loves these animals more for its own sake than
for theirs, *i.e.*, for the pleasure they afford him.   It does not
seem to occur to him that these creatures can also suffer
and enjoy.   Their gambols, their happy bark or mew, delight him ; their cries of anger or pain frighten and distress
him, and that is about all.   The dog being more good-
natured than the cat, there is no trial to which this child
will not subject its patience; I have seen him pull him
by the tail, by the paws, by the ears, and even bite his
tail, thrust all sorts of things into his jaws, throw all his toys
on him, drop a chair down upon him, or beat him with his
wooden spade, etc., etc.   The other day, the nurse seated
this child on the lawn, and put a tortoise near him to amuse
him.   At first he looked at the animal with great curiosity,
and seeing this, the nurse left him alone for a moment ; on
her return, the tortoise had one leg half torn off, and this
enthusiastic student of natural history was occupied in pulling off another with all his strength.   This insensibility to
the suffering of animals, unless it is very evident to the eye
or ear, is very common in children, and even in a great
number of adults ; but it is due to a faulty education, rather
than to a defect in natural sensibility.   I have often heard
ignorant, uneducated, people assert that such and such animals or insects could not feel pain.   And have not the extreme advocates of the automatism of beasts,—Malebranche
among others,—positively declared that they do not feel?
It is remarkable, moreover, that a learned man, the celebrated Lamarck, one of the precursors of Darwin, has distinguished a part of the animal kingdom by the name of
*apathetic* animals.

*Human Sympathy.*—A child of twelve months who came
back to his father's house after a month's absence, took no
notice of the purrings and caresses with which his old friend

the cat welcomed him home. He hardly noticed the dog either, though he had been in the habit of seeing him every day, and had sometimes been allowed to play with him, and used to repeat his name. It took him ten minutes or more to recover his familiarity with either of them. Scarcely however, did he catch sight of the faithful old servant—before even she had called him by his name—than he held out his arms to her, starting and jumping with delight. The fact is, that though children often seem to love cats and dogs as much as their parents or nurse, they forget animals much faster than people. A rather older child remembers animals much better, and will even speak of them constantly when away from them. But I think that in general the affection they feel for people is of a deeper nature. Setting aside the tendencies to human sympathy which are the result of organization and heredity, man is always necessary to man ; and this is especially the case with the little child, that "*être ondoyant et divers*," the subject of ever new curiosity, always seeking fresh gratification. A little child hangs, in the full sense of the word, on the looks, words, and actions of the human beings around him. Human speech is in itself, apart from the ideas and sentiments it expresses, a music, of which the rhythm and the intonations correspond to the æsthetic faculties of a little child. The ever-changing expression of the eye lends a further charm to this delightful music. The eye is one of the most interesting and attractive of objects ; the vivacity of the pupil set in its oval background of white, its sparkles, its darts of light, its tender looks, its liquid depths, attract and fascinate a young child, like the roundness and the shifting patterns of a beautiful agate ; it is a source of perpetual enchantment. Although children may not be subtly sensible to all the delicate impressions produced by beauty, graceful figures, pleasant faces and manners, cannot but have a powerful attraction for them. The play of the human countenance, moreover, so strongly affects the organization of sensitive beings, that even animals who do not live habitually in the company of human beings will endeavour to find out its meaning. And all these objects of intense and incessant curiosity are associated in a child's mind with the ideas of caresses, games, affection, pleasure, nourishment, *i.e.*, all the pleasures, moral, intellectual, and physical, that they have experienced.

One sees in children, especially towards the age of ten months, sudden fancies for new faces which it is not always easy to explain. A young relation of mine, eleven months old, came once on a visit to my family. On first seeing me, his father being present at the time, he called me, "*Papa, papa.*" I held out my hand to him and he seized hold of my fingers and dragged me along, saying : "*Papa, mené, mené.*" I at once understood that he wanted me to help him to walk, and to his great delight I gratified his wish. During the whole week that he remained with us he continued to show a marked preference for me over every one else. He would often leave his mother, and his father still oftener, to come to me. Often at meals, when he was on his mother's lap at a considerable distance from me, he would fix his eyes on me, say "*Papa, papa,*" and then slip from his mother's lap and scramble on all fours under the table up to my feet, in the hopes that I should take him up. Sudden affection of this sort deserved to be examined into, and I believe that I found the key to the mystery. On the day of his arrival, seeing me smoking a cigar, he began to puff vigorously as if he were blowing smoke through his lips. Now he goes through this performance with his grandfather ; the latter, moreover, whom he also calls *papa*, has a long beard like mine. It was no doubt these points of resemblance, and possibly some others, such as likeness in manner, figure, and voice, which caused me to become at once the depository of the affection before bestowed on his grandfather. Now, however, he has learnt to like me for myself, or rather, I should say, because of the games I have played with him, and which he cannot enjoy without me.

Human beings are thus the objects of children's most marked affections, as they seem also to be of tame animals. It is seldom, however, that a child manifests its sentiments with a sufficient degree of energy to merit the name of passion. Exalted love—so-called passion—is the characteristic of the adult. It implies an element of reflection, though not perhaps always reasonable, and a strong impulse of the will, though not always regulated ; and these elements are wanting in young children. They have no such things as passions, but, like animals, they have attachments and habits. As regards sympathy properly so called, we must not look for anything more in them than the germ of this sentiment. A young child has not yet made sufficient trial of the good and evil of life, to be able to imagine them in his fellow-creatures. He does not sufficiently understand the full signification of the facts which he observes, he does not possess in a sufficiently high degree the faculty of induction and judgment for his sensibility to be affected by the external manifestations of complex sentiments. He does not experience those moral sufferings which, for the adult, are often far harder to bear than any physical pain. He may suffer from the deprivation of a beloved person or object, but he does not say to himself that he suffers. Suffering is only connected in his mind with tears and groans.

This period of life, so full of irreflective sympathy, is, in the words of the fable, without pity, owing to want of experience and to feebleness of judgment. Every day we see children, even as old as three or four years, innocently doing violence, by their inopportune remarks and cruel proposals, to the most sacred griefs of those who love them. I remember that, when about five years old, having lost a young sister, I was taken by my aunt to the bed where the dead child was lying. The pallor and immobility of her face, her half-closed eyes, and her distorted mouth (she had died of croup) all made a deep impression on me ; but at the same time, she reminded me so strongly of a very pale-faced little boy, whom I had often noticed on my way to school on account of his colourless complexion, that I could not rest till I had found my mother to tell her of this likeness. A child of four years old had lost one of his favourite companions ; he was taken to the little boy's house, and the father took him on his lap and held him there a few minutes while giving way to a fit of weeping. The child

understood nothing of all this grief; he got down as quickly
as he could, disported himself a little while about the room,
and then suddenly going back to the poor father, exclaimed :
"Now that Peter is dead, you will give me his horse and
his drum, won't you?"

Sometimes, however, we see exceptions to this rule.   I
know a little child, not yet three, who is a striking
example of what has been aptly called the "memory of the
heart." His grandfather and father, whom he was very fond
of, are both dead. The grandfather has been dead a year,
and the father five months. Not a day passes that he does
not speak of them. Whenever he is taken to his grandmother's house, he seats himself on his grandfather's arm-
chair, asks to have the curtain lifted up from grandpapa's
picture, and looks at it with an expression of real emotion.
At home he asks every day for the photograph of his father;
he kisses it and makes his sister and his elder brother do the
same. He understands why his mother is sad, and says to
her, "You are sad because father is not here." If he sees
her crying, he kisses her and says, "Don't cry; I will go
and fetch papa, I will make him come back; I have got
the key of Paradise."

Let me mention one or two more facts to show how
very unequally different kinds of sympathy may be developed
in the same child. The following facts were furnished me
by a friend : "Since I have been under hydropathic treatment, it has twice happened that the child (sixteen months
old) has been present at the operations. . . . Each
time, when I began to douche myself, he burst into tears,
probably, I suppose, because he remembered what a disagreeable sensation he feels himself when he is put into a
bath. The first time, I was obliged to leave off the performance and put him out of the room away from this
painful sight. The second time, I let him remain, but he
cried the whole time. He fetched my clothes from the
chair, and held up my shirt for me to put on. This sympathetic sensibility touched me deeply."

"The little fellow is perfectly miserable if either his
mother or I say to him, 'I am angry, baby.' If any one

scolds him, it is the displeased expression of face that causes him the **most distress and sets him off** crying. If the scolding is not of a very decided nature, one sees him **hesitate,** his mouth is uncertain whether **to** laugh **or** cry, and he finally decides according to the dominant expression. If he cries a great deal **at** being scolded, his mother can always comfort him **by saying,** 'Mother not cross now; be mother's **little darling,' and then** he **is instantly** consoled and holds **up his face to be kissed."**

But the reverse of the medal is always there also. This same child at the same age, and **even a year later, was the** terror of all cats. During a visit that he paid at my house, I went one day suddenly into **a** room where he had been left alone with a little kitten. On seeing me he cried out, " I'm not hurting **the kitten."** This was true at the moment **at** which he spoke, for **I** found the little creature squatting **under a cupboard,** frightened to death.

# CHAPTER VI.

## INTELLECTUAL TENDENCIES.

A CHILD's curiosity, like its intelligence, is at the outset entirely egotistic and sensual, but it is gradually and instinctively elevated by a kind of scientific disinterestedness. It is in the first instance a vivid excitation of the sensibility, and, by repercussion, of the activity, in the presence or expectation of new and vivid sensations.

Fénélon has given a psychological, if not an exactly physiological description, of this interesting organ of the infant mind. "The substance of their brains is soft, but it hardens day by day. As for their minds, they know nothing, all is new to them; owing to the softness of the brain, impressions are easily traced on it, and the novelty of everything causes them to be easily excited to admiration and curiosity. This moisture and softness of the brain, combined with excessive heat, produce in it easy and constant movement; hence arises the restlessness of children, and their inability to keep their minds fixed on one object, any more than their bodies in one place." Fénélon says elsewhere : "The curiosity of children is a natural tendency which goes in the van of instruction." Let us say rather, "in the van of pleasure," and the definition will be more exact.

At two months a child will turn its eyes and ears and stretch out its hands and arms towards the objects which strike its senses. At three months it seizes the objects brought within its reach and shakes them about to amuse itself; it knows that its hands are the instruments by means of which it can procure itself impressions and produce movements, and it exercises them in touching and in bringing near to its eyes, and better still its mouth, as many objects as it can. When it is undressed it rubs its hands all over its little body, down its stomach, along its legs and feet; and it is astonished to feel so many different things which are all parts of itself. When it is seated on its nurse's lap or on a cushion in the middle of the room, one of its favourite occupations is to take its foot in both hands and to pull it up to its mouth. The mouth is the central point of all its young experience; it is here that every fresh bit of know-ledge is brought to be measured and determined. Very often too, whether from the need of testing everything by the taste, or from the desire to sooth the smarting of its teeth or gums, and to arrive at the remedy by what seems the shortest road, children will try to take hold of distant objects with their mouths.

Then soon comes the stage at which everything within reach becomes the subject of continual study and desire, and the child's curiosity flits from one thing to another and backwards and forwards to the same things over and over again, changing as rapidly as does the pleasure which he feels in holding, moving, looking at, and listening to different things. Woe henceforth to you parents who have not kept your children in check, but have made yourselves the ready slaves to their caprices. By a graceful wave of the baby hand, or by resolute and imperious screams, they will now insist on your bringing or giving them whatever strikes their fancy; your watch, your eye-glass, your arm-chair, a picture, a porcelain vase, a lamp, possibly even a gas-burner in the street, "everything," as Rousseau says, "including the moon."

Towards the age of a year, when a child begins to be able to walk, its sphere of personal investigations becomes rapidly enlarged, and the additional faculty of speech supplies its curiosities and wishes with the means of endless variety, and of enforcing attention. A hundred times an hour, provided there is some one to listen to him, his little voice will be heard, expressing some wish or asking some question. All the observations which he formerly made with his eyes, are now made with his hands and mouth;

he darts about hither and thither, he crawls or toddles from
one thing to another ; he opens things, breaks them, knocks
them about, mixes them together.   He pours his broth into
his grandfather's watch, puts the gold fish into the doll's
bed, and the doll into the water of the fish-globe ; in short,
he commits the whole series of incongruities which artists
of late have been ingeniously striving to reproduce ; and all
this is done, not so much from a desire to know what the
things are and what can be done with them, as from the
need of fresh sensations.

A little later, these mischievous tendencies become still
more numerous.   The child seems to be everywhere at
once—in the kitchen, in the garden, in the drawing-room,
with eyes and ears wide awake, hearing and seeing everything without seeming to do so, asking endless questions,
often very embarrassing ones, and storing up in his memory
all the most striking details, to be brought out suddenly for
the entertainment of visitors.   This craving of young children
for information is an emotional and intellectual absorbing
power, as dominant as the appetite of nutrition, and equally
needing to be watched over and regulated.

We should strangely exaggerate the hereditary influence
of scientific tendencies in man, if we transferred it from the
social group to the individual : all that belongs to the individual is a greater aptitude at perceiving and at combining
his perceptions so as to form systematic and co-ordinated
conceptions; but the instinct for abstract truth, the necessity of finding out the truth for its own sake, is not transmitted, it must be inculcated.   If then there are no questions
asked by children to which a true answer should not be
given, we may comfort ourselves with the thought that little
children are not very critical, and that a very vague answer,
if necessary, will satisfy them.   For instance, a little girl
asked her mother why there was water in the river.   Her
mother answered, " Because there must be water somewhere
but not everywhere."   Another child asked why beans
grew in the earth.   Her mother answered, " Do you not
grow every day? and kittens, don't they grow ?   All animals
get bigger, and little ones become great ones ; and the

plants do just the same." **Another child asked** why water
was not wine. His father asked him in answer, " Is a dog
a cat? Wine is wine, and water is water." All these
answers are not partly true, they are absolutely true, and
quite sufficient for children of this age.

Children do not trouble themselves about the invisible;
they feel, indeed, an instinctive repulsion for what is mysterious. It may be said of them, as Spencer has well said
of primitive man : " He accepts what he sees, as animals
do ; he adapts himself spontaneously to the world which
surrounds him ; astonishment is beyond him." If, then, it
is not possible to attribute the birth of religions to those
two tendencies said to be innate in man,—automorphism,
which causes him to place a will similar to his own behind
natural phenomena, and wonder, which seizes him in the
presence of certain of these phenomena and impels him to
invent mysterious and supernatural explanations for them,—
we may boldly assert that the sense of religion exists no
more in the intelligence of a little child, than does the
supernatural in nature.

Children, I take it, occupy their minds very little with
those conceptions of the invisible, of the infinite, and of
finality, which are disputed about in philosophic circles.
Their reverence and their love attaches itself to the human
beings who are kind to them, but to nothing which is invisible or distinct from their species. Their instinct of
finality is wholly objective and utilitarian. They ask what
such a thing is called, in order to know in what it is either
good or bad ; and also—but this chiefly because they have
been taught to ask it—where it comes from, who has made
it thus, who has put it there ; *i.e.*, what is there to like or
fear about any particular thing? There is nothing metaphysical in all this ; they are only inquiries founded on
very concrete analogies and experiences. The mystery of
their own existence and of the existence of the world does
not interest or preoccupy young children, unless they have
had their attention directed to these subjects ; and, in our
opinion, parents are very much mistaken in thinking it their
duty to instruct their little ones in such things, which have

no real interest for them—as who made them, who created the world, what is the soul, what is its present and future destiny, and so forth.

So great, moreover, is the confidence and credulity of children, that they will accept,—not always though without a little kicking,—almost any beliefs that you try to impose on them. If you want to persuade your child that he was born under a cabbage, that Hop-o'-my-Thumb had seven-leagued boots, that the sky is peopled with angels, that under the earth there are howling demons, that garrets and chimneys are full of ghosts, you have only to look as if you believed all this seriously yourself, and they will be convinced at once.

## II.

### VERACITY.

Montaigne has said that the falsehood and stubbornness of a child grows with its growth ; and with regard to the first of these faults, of which I shall treat in this chapter, he has spoken very truly. Like all hereditary vices, the habit of lying is developed more or less in each child, according as internal or external circumstances, education, example, the influence of such and such other tendencies or habits favour or counteract its growth. As a matter of fact, it is an absolutely subordinate vice, and we must search out its primary source and its accidental derivations if we wish to apply a timely and efficacious remedy to the evil.

A child's truthfulness is in proportion to its credulity, although there is no direct relationship between these two qualities. Nevertheless, at a very early age the illusions from which no one of its senses is exempt, begin to startle its early confidence, though without at once shattering it. There is no doubt that many things astonish children on account of the distinction they are obliged to make between them and other things which resemble them. I have seen a child of five months old quite perplexed and bewildered at discovering that what she had taken for one and the same cat was two cats of the same colour. Very young children are frequently surprised and irritated by mistakes which they have fallen into. But they are much more annoyed and irritated by deceptions practised on them by others.

If you give a child of seven or eight months a piece of bread instead of a cake which he had seen and coveted, he will thrust aside the mocking gift with a gesture of disgust; and the twitchings of his mouth, his tearful eyes, and puckered forehead threaten an imminent outburst. Who that has had to do with children is not familiar with the terrible scenes that occur when a change of nurses, or the necessity of weaning the child, obliges those around it to resort to all sorts of little tricks and manœuvres. Happy the parents who can flatter themselves that neither they nor any of their household have ever deceived a child unnecessarily, and without legitimate cause.

Thus then we see that children from twelve to fifteen months old have already discovered that all that grown-up people do or say is not always truth. Very soon, moreover, if the phenomenon of falsehood has not already come under their notice, the cunning which is innate in every animal organization will lead them to it by the practice of useful dissimulations. The story of the piece of sugar stolen by the young Tiedemann or the young Darwin, is the story of all children at the same age.[1] They hide themselves instinctively to do what they know is forbidden, as if in play, just as they will say what is not the case by way of fun. A child of two years who says to me : "I have just seen a butterfly as large a cat, as large as the house," is telling for fun what he knows to be a falsehood. It is the same, too, when he crouches behind a door saying, "Victor isn't here." But of these two untruths the one is spontaneous, the other is imitated; the playful imagination of children, and their tendency to imitate the games of others, are two sorts of inducements to them to counterfeit the truth. But in whatever way they do it, whether by gesture, mien, or speech, it is more for their own edification than

[1] See my brochure : *Thierri Tiedemann et la Science de l'Enfant*, etc., p. 36.

for that of others. They enjoy the surprise or the fright which they think they have caused their nurse by suddenly popping their head out from behind their table-napkin, or by coming out of a hiding-place where they thought themselves invisible. "How I frightened you!" said a little girl of twenty months to her uncle, who had pretended to be frightened at hearing her imitate the barking of a dog behind the door. We see also from example that the gratification of the feeling of *amour-propre* has something to do with this tendency of deceiving for fun.

Children will also turn a thing into a joke in the same sort of way, to avoid being scolded, or to appear as if they did not deserve a scolding. "Naughty, naughty," said a little child of two-and-a-half to its mother who was putting it into a bath against its will. "What!" said the mother, "is that how you speak to me?" "No, no, it's not you; it's the water that's naughty," was the prompt reply. In this case the falsehood, though suggested by the mother's question, had more the nature of spontaneity than imitation.

All egotistical feelings are conducive to lying. A child who has just had some good thing to eat, will say that he has not had it, or that he has had very little, in order that he may have some more given him. Another child has burnt his mouth, say, in drinking his broth, and begins to cry. His nurse tries to comfort him, saying, "Poor baby, it hurts very much, doesn't it," etc., etc. All these words of comfort the child repels with movements of the arms and head, accompanied by unintelligible sounds, till at last it says more distinctly: "No, no; not hurt."

In this case the imagination, over-excited by pain and anger, suggests to the child the idea of denying the reality, because he does not wish to believe it. A little girl, three years old, seeing her mother fondling her brother for several minutes without taking any notice of her, said all of a sudden: "You don't know, mamma, how naughty Henry has been to the parrot." This was a falsehood suggested by jealousy.

Laziness will also lead children to tell untruths. A child is told to go and look for a stool in the next room; he

returns without having looked for it, and says it is not
there. Or he is given a book to take to his uncle who is
sitting out in the garden. He reluctantly leaves his playthings, hesitates before starting, walks as slowly as he can,
looking round several times to see that no one is looking at
him, and when out of sight, he drops the book into a bed
of flowers, and runs back to his toys, as if he had executed
the commission. Laziness, disobedience, and hypocrisy are
all combined in this action.

The fear of being scolded, or punished, or of being
deprived of an expected treat, will also often lead a child
into falsehood; but this is generally at an older age—after
they are three or four years old. Very often, moreover, it
is our manner of trying to elicit the truth which leads to
defiance, dissimulation, and falsehood. Did *you* do that?
*Who* has done this? Questions like these, asked in a
threatening voice and with severe looks, provoke an answer
which may save from punishment, or at any rate may put it
off; and so the child tells a lie.

Truthfulness is so essential a virtue, and the habit of
lying is so dangerous, and one which in so many ways
affects all the details of human life, that we cannot be too
much on our guard against the first symptoms of prevari-
cation. It is a matter in which preservatives are more
valuable than correctives. It is a common-place truth, but
one too much forgotten in actual practice, that, with justice
and kindness, we can make almost anything we like of
children. They will be frank and open, if they are encouraged to be confiding; they will not seek to make
excuses for their faults, if they know that it grieves their
parents, and that this will be the worst consequence of
their naughtiness. And lastly, we should be specially careful never to scold them for unintentional faults.

### III.

#### IMITATION.

During the first months of life, the imitative tendency is as
little developed as the power of observation; it exhibits

itself, however, in various little efforts. Darwin thinks that he noticed it in his son at four months old, when the child appeared to imitate certain sounds; but he only determined it positively at the age of six months. Tiedemann has noted the following phenomenon in his son at four months, and he thinks it a clear indication of association of ideas: " If he sees any one drinking, he makes a movement with his mouth as if he were tasting something." In this and analogous actions must we not recognise, besides the instinct of finality which makes the child understand the object of buccal movements, the result of that instinctive sympathy of movement which, in beings provided with the same organisation, makes like call forth like, and, given the impulsive character of childhood, results naturally in imitation?

M. Egger,[1] while agreeing that the faculty of imitation is very precocious in children, does not note it with certainty until the age of nine months, and then in the following actions:— 1. Alternately hiding and showing themselves, as a game. 2. Throwing a ball, after having seen some one else throw one. 3. Trying to blow out a candle. 4. Trying to sneeze, in imitation of some one else. 5. Trying to strike the keys of a piano. M. Egger does not notice at the same period "any conscious effort at imitating sounds." He considers, moreover, that the development, or rather the appearance of this faculty of imitation, is simultaneous with the first awakening of intelligence. For my part, admitting as I do the co-existence in children at a very early age of automatic movements and conscious and intelligent movements, I believe that the imitative function is also very early connected with these two sorts of movements.

Owing to the nervous connection of the auditory and phonetic apparatus, young birds attempt mechanically to produce the notes they hear sung by adult birds; talking birds try mechanically to reproduce the words and the sounds which strike on their ears; and mechanically also,

---

[2] *Le Développement de l'Intelligence et du Langage chez les Enfants,* pp. 10, etc.

as it seems to me, children, as soon as the second month, make attempts at sounds, in order to answer people speaking to them, or to imitate the sounds, of the piano. Likewise at the age of three months, having learnt mechanically to follow the direction of a look or a movement, they will fix their eyes on objects which they see other people looking at, turning their head also sometimes where they see other persons turn *theirs*, and all this, though in a very limited measure, with the intention of imitating. The same applies to all young animals.

As the simultaneous development of the faculty of observation and of the play of the organs proceeds, the sphere of children's imitations is proportionately enlarged : for instance, their arms at first attempted a variety of instinctive movements towards objects which attracted their attention or excited their desires ; the desire to imitate analogous gestures, which he sees others make with success, excites him to renew his efforts, and indicates to him the way to succeed in his turn : at the age of four months he stretches out his arms with more assurance to the people around him ; he is more successful in smiling, he even attempts to laugh —and all this improvement is due to the attempts at imitation which have helped and strengthened the first spontaneous efforts. At six months children respond with little starts and jumps to any attempt to amuse them ; they stroke their mother's face with their little hands, they babble inarticulate expressions of admiration at any object which they are made to contemplate. A little later still, their *m m m* and *p p p*, more or less spontaneous, becomes the *mamma mamma*, and *papa papa*, which have been repeated to them a hundred times a day. And finally their first efforts at walking, partly due to their own initiative, and partly to the guidance and help of nurses and mothers, will be brought gradually to perfection through their anxiety to imitate the grown-up people whose walking they watch so attentively.

The older a child grows, the more he will be helped by example in guiding the operations of his senses, as well as those of his moral and intellectual faculties. For instance, as has been said above, "We feel the desire to eat arising, when we see others eating," and the practical lesson to be learnt from this is, not to give the stimulus of example to the instinct of greediness innate in children. The same may be said of all actions which have more or less connection with sympathy, sociability, or with anti-social tendencies. A little girl only fifteen months old had already begun to imitate her father's frowns and irritable ways and angry voice; and very soon after she learnt to use his expressions of anger and impatience. When three years old, this same little girl gravely said to a visitor at the house with whom she had begun to argue quite in her father's style : " Do be quiet, will you, you never let me finish my sentences ! " Thus we see that the contagious effect of example operates very early on the habits and morals of children. They copy everything, evil as well as good, and are very quick to adopt the opinions and ways of acting of their elders. We must not, however, be in too great a hurry to judge them from their passing estimate of particular actions, or from their imitation of them, which is purely accidental. It is not so much habits as tendencies to habits which they exhibit, and these are often lost as soon as learnt, as well as the manners or language which they have acquired in the same fashion. Moreover, there is in every child a distinct individuality, the result of heredity and habits, which can always be discovered if we will but look for it, underneath all its plagiarisms.

The spontaneity natural to early infancy is sometimes the means of saving children from the inconvenient results of their extreme organic and intellectual plasticity. But it would be dangerous to count too much on this spontaneity. The respect due to the individuality of a human being makes it incumbent on us to be very careful as to the examples a child sees around him, especially from the moral point of view. The ideal in education would be, to allow each child scope for his own particular bent, while at the same time setting our example before him. Locke understood the necessity of respecting the natural bias in each child, and could not endure the artificial product which is the invariable result of constraint and affectation. He specially deplores this fault in what concerns manners and behaviour in society. "Affectation," he says, "is a clumsy and forced imitation of what should be easy and natural, and is devoid of the charm which always accompanies what is really natural, because of the opposition which it causes between the outward action and the inward motions of the spirit. . . ." Away with politeness and agreeable manners if they endanger the frankness and sincerity of the child. "Mamma," said a child of four years old, "are you not going to tell Madame X. to go away? She has been here a long time." I greatly prefer, even in a child of four years old, this frank and innocent rudeness, to formulas of politeness repeated by rote but not felt.

There is another reason, well worth our consideration, which should deter us from stifling a child's natural initiative by the undue influence of our example and activity. We see in animals a sort of individuality of action which does not belong to man; the development of their powers and skill affords them the greatest possible amount of enjoyment when they are young, and later on inspires them with a kind of proud confidence. And the same thing may be observed in little children. Tiedemann says of his son at fifteen months old,—and the observation might have been made earlier,—"When he has done something of his own accord, given a certain impetus to one of his toys, for instance, he shows evident delight, and takes pleasure in reiterating the action." And he goes on to remark with equal truth: "Children in general like to do by themselves what they have hitherto been obliged to let others do for them. They like to feed themselves with their own hands, to wash and dress themselves, etc., etc. This liberty of action, even in imitated actions, is one of the conditions of a child's happiness; besides that, it has the effect of exercising and developing all its faculties. Example is the first tutor, and Liberty the second in the order of evolution; but the second is the better one, for it has Inclination for its assistant."

## IV.

### CREDULITY.

The instinct of credulity is at first nothing else than the instinct of belief. It is impossible for us to have a sensation, or a perception, or in fact a clear idea of any sort without referring it at first to some object which is present, and later on to something which has been or might be present—in short, to some real and tangible cause. This is why, in the case of children or young animals without discernment, and, up to a certain point, in the half-civilized adults whom we call savages, belief, or, if we prefer it, credulity, is confounded with the desire to see such and such objects either present to the sight or fixed in the memory, and to see them with such and such attributes, agreeable or disagreeable. The child begins with a lively faith in the truth of appearances ; but day by day he goes through experiences which teach him to mistrust, not appearances in general, but certain sorts of appearances. In this respect his faculty of discrimination and generalization makes marked progress during the important period from the third to the sixth month.

A little child of seven months who is on a visit to me, is seated on the table before me as I am writing. I place a brush within his reach with the bristles turned towards him : he presses both his hands upon it, but soon lifts them up again—very slowly, and looking very grave. His attention is then attracted in another direction. Some minutes after, I try the experiment over again, and this time I notice a little more rapidity in the child's recoiling movements. I repeat it five times more at intervals, and varying the circumstances ; but I do not remark any new facts. A quarter of an hour having passed after the seventh experiment, I again place Georgie in a position to touch the brush. This time he draws back quickly at first sight, and without trying to touch it. Having then amused him and distracted his attention for a short space of time, I repeated the experiment once more for the last time. The child looked fixedly at the brush without stirring, then, after a few minutes of hesitation or reflection, he threw himself back and kissed his grandmother.

Children make experiments of this kind for themselves every day; and though they *generalize* them very little, they help them,—by dint of repetition, and thanks to analogy and association of ideas,—to know how to act advantageously with regard to new objects which come before them. But the mistakes which they make so frequently they do not realize as such; they regard them merely as isolated experiences; and thus in most cases the little creatures, irreflective rather than inexperienced, judge things from their appearance.

*A fortiori* then will children trust blindly in the words of others. " The sounds which a child has become accustomed to recognise and to attach to certain objects while learning to speak, will naturally awaken in him the thought of the same objects when heard again : until these associations have been disturbed by the experience of error and falsehood, they reproduce themselves naturally and infallibly ; the same words recall always the same ideas. If every time that one says to a child, 'I have a cake for you,' one really gives him a cake, it is impossible but that the same words, spoken another time, should not awaken in him the same idea and expectation ; but if once, instead of the promised cake, he is given a bit of wood, he finds himself suddenly confronted with a falsehood that nothing could have made him foresee, and he begins to cry. We can plainly see that the instinct of veracity is not needed to explain these facts." I am of the same opinion as M. Janet : it is not necessary to have recourse, as Reid has done, to the instinct of truth, to explain the natural belief of human beings in the testimony of other human beings. This fact has its principle in the natural belief of a child in the sense expressed by words, that is in the objectivity of the ideas that words express. Even at the age of two or three, when the child has made experience of error and falsehood, he will tell untruths himself, without ceasing to believe in the veracity of others. He only mistrusts certain people, and those never entirely. Untruthful and credulous, these are two qualities which go very frequently together in childhood, and also, I think, in adult life ; it is easier to tell lies than to believe that others are doing so.

Children, then, accept unhesitatingly as true all the ideas which pass through their brains, and especially those which gain confirmation and precision from the words or looks of grown-up persons. The tales of the latter, whatever they may be, instantly become the child's creeds. This is what constitutes the charm, and also, in my mind, the danger, of all the improbable stories which are related to children, before even they thoroughly understand the language in which they are told.

At the age of twenty months a child is not keen to hear stories and fables, which he would not understand; but he delights in recounting his own little experiences. A little girl of this age, whenever her mother took her out with her, used to relate to her father in the evening all that she and her mother had seen and done: "We went out under the large trees of the Luxembourg; the dog was with us; he kept running round the perambulator of a little girl, and every now and then he came up and licked her hands and face. But the dog was very naughty, he ate the little girl's cake. Mamma scolded the dog well, and drove him away with her blue umbrella, which made Mary laugh just when she was beginning to cry. Then a little boy named Joseph came and sat on a bench by Mary. He was bigger than little Mary, but he was very polite, and he is very fond of the little girl. He let her take his balloon and he did not hurt her doll; then he and Mary jumped about together, but the little boy tumbled down and made a bump on his forehead. He cried very much, and the little girl cried too because he was hurt. And then we walked a long, long way to the furthest bench with Madame X., who loves baby very much. Madame X. said to baby: 'When are you coming to see me? There are some beautiful apricots for you in the garden, and the birds in the aviary are always very pretty and very happy: they often ask where little Mary is, saying, *Coui, coui, coui*,' etc., etc." And during this recital, often interrupted by the kisses and pettings of her mother, or by bursts of laughter and short remarks from her father, the little girl, all eyes and ears, enacted all the various emotions which the events called forth, gesticulating with arms, feet, and head, and mimicking the cries of the animals she was talking about. She would become half lost in the narrative, or rather in dramatizing it ; and the habit of recounting these true stories prepared her for following the fictitious ones which her mother invented for her, suiting them gradually to the progressive development of her intelligence. When two years old, she could not exist without these exciting little tales ; and she used to say several times a day to her mother, " Mamma, tale about dood ittle dal ; mamma, tale about ittle dal."

These little dramas and comedies, of which children are so fond, are taken by them quite seriously, even when they get to the age of three. When once they know them by heart (a particularly happy expression in their case !), if you change a single word, the charm is gone, and the attention with it. A child who went on a visit to some relations when only two-and-a-half years old, was particularly fascinated by the tales which his youngest aunt used to tell him. He only liked to hear them from her, and he wanted them over and over again every day. Often after late dinner, if he did not fall asleep in the middle of dessert, he would settle himself comfortably on her lap, leaning his head on her shoulder, and remain perfectly still waiting for her to begin. And then nothing could make him stir ; no noise or interruption would disturb his immovable attention. But we could tell by the expression of his face, by his sudden changes of colour, and by the movements of his eyes and the twitchings of his lips, what a series of profound emotions was successively agitating his little mind.

One evening his aunts were out, and he was left alone with the eldest of his cousins, a young man. He soon began to find himself bored. His cousin proposed to tell him the story which he liked best, the one about a young bird, which, having left its nest, although its mother had forbidden it to do so, flew to the top of a chimney, fell down the flue into the fire, and died a victim to his disobedience.

The narrator thought it necessary to embellish the tale from his own imagination. " That's not right," said the child at the first change which was made, "the mother said

H

this and did that." His cousin, not remembering the story word for word, was obliged to have recourse to invention to fill up gaps. But the child could not stand it. He slid down from his cousin's knees, and with tears in his eyes, and indignant gestures, exclaimed, " It's not true ! The little bird said, *Coui, coui, coui, coui* before he fell into the fire, to make his mother hear ; but the mother did not hear him, and he burnt his wings, his claws, and his beak, and he died, poor little bird." And the child ran away, crying as if he had been beaten. He had been worse than beaten, he had been deceived, or at least he thought so ; his story had been spoiled by being altered. So seriously do children for a long time take fiction for reality.

# CHAPTER VII.

## THE WILL.

I KNOW not if it be very daring to suppose that all the movements produced during uterine life are not absolutely unconscious, and that the child may have some sort of idea of a few of them at the moment they are going to be produced. For instance, we may ask ourselves if it has in no measure preserved the recollection of certain sensations of temperature, of pressure, of contact, of taste, and also of movements more or less defined, such as contortions, tremblings, contractions of the muscles of the mouth, which the organism will have made in response to these different sensations. If it were so, the consciousness which, according to our hypothesis, would be either concomitant with or consecutive to these instinctive movements and the sensations which determined them, would tend to forestall the action : or it might diverge into two separate actions, bearing successively on the sensations and on the movements in question. This is how it would be : a sensation would be strong enough to be felt, but not strong enough to produce instantaneously the corresponding movements ; this hesitation of the instinct of movement in following the shock of the sensation,—perhaps a pause produced by two crossing sensations,—would give the child time to interpose a glimmer of consciousness between the sensation and the action ; and this would be a first slight beginning of spontaneity. Is it at any rate altogether exaggerated to suppose that during the long travail of birth, composed of such varied actions and interludes, the child, however little conscious of some of the efforts which it makes itself, and of the sensations of different

degrees of intensity which cause it to make them, is nevertheless capable of accidentally producing some of these
movements with a vague desire of doing so? It is this intervention of consciousness and desire (or of personal spontaneity in the production of movements fatally determined),
if it does exist before and during birth, that we must look
upon as the germ of voluntary activity.

But without troubling ourselves further as to this metaphysical hypothesis, let us try to interpret scientifically some
few of the facts observed at a later date. That which distinguishes voluntary movements and actions from involuntary
ones, is, that in the former there exists a desire either to
renew or to get rid of certain sensations, and besides, some
idea of the movements necessary to produce this result ; in
other words, some idea of the muscular sensations corresponding to these movements. The desire may be more or
less vivid, the idea of the movements more or less distinct,
and by so much is the intervention of the will greater or
less. It is generally very slight during the first period of
existence. Whether the great nerve centres, motor and
other, be incomplete at birth, whether instinct makes easy
to the child a large number of the movements which it seems
to learn, the volitions,—that is to say, consciousness, effort,
and desire,—play very little part in those movements which
are predetermined by organization and heredity. Thus
the fingers of a child a few days old close too easily on
any object placed in contact with its palm, for it to be very
early stimulated to produce similar movements voluntarily.
In like manner it may be maintained that the progress made
in the act of suction during the first days has gone on with
as little of consciousness and desire as possible, that this
progress is due to a development of forces and to exercise,
which are purely mechanical; but can the same be said
*apropos* of the remark made by Tiedemann on his sixteen-
days-old child. " If put in a position to suck, or if he felt
a soft hand on his face, he left off crying, and felt about for
the breast." There appears to be something more here than
association of ideas. There seems to be a recollection of
and a desire for the action of sucking, and perhaps for the

movements of suction.   M. Preyer only recognises the first
appearance of volition in a child in its holding its head upright, and this it does not do till the end of fourteen weeks.
We may, however, notice other conscious efforts towards
the same period.   When a month and five days old, Tiedemann's child " distinguished by himself objects outside him,
showing the first effort to seize something by extending his
hands and bending his whole body."

The same movements, more or less conscious, are to be
remarked in cats and dogs before the end of the first week.
The little Tiedemann, however, seems to me much too precocious.   The following observation has much more veri-
similitude.   When a month and twenty-seven days old, the
same child already showed that he realized his own
activity ; his gestures of pleasure indicated this, as well as
his fits of anger and violence in pushing away disagreeable
things.   The remark made at the same period by Tiedemann,
on the imperative intention of tears, is confirmed by similar
observations of Charles Darwin's.   At the age of eleven
weeks, in the case of one of his children, a little sooner in
another, the nature of their crying changed, " according to
whether it was produced by hunger or suffering."   This
means of communication appeared to be very early placed
at the service of the will.   The child seemed to have learnt
to cry when he wished, and to contract his features according to the occasion, so as to make known that he wanted
something." [1]   This development of the will takes place
towards the end of the third month.   At the same period,
Tiedemann's son was still at the stage of those instinctive, or rather primitive, movements of desire which show
themselves by the mechanical extension of the arm and
the inclination of the whole body ; it was not until
towards the fourth month that he began to take hold of
objects placed within his reach ; and in carrying them to

---

[1] Examples taken from Darwin's " Expression of the Emotions,"
and which I have reproduced in my brochure : *Thierri Tiedemann et
la Science de l'Enfant.   Mes Deux Chats*, a fragment of comparative
psychology, pp. 12 etc.

his mouth he learnt to hold them firmly; all these steps in advance being conscious and intended, that is to say, voluntary. In proportion as the understanding and the motive forces, or, in other words, the great nerve centres of the child, develop, the emotional element of volition or desire, and the idea or conception of the movement to be made, become more marked. Desire is so much the more keen, as the conception of the object desired becomes clearer. This is the case with all movements, which, though partly instinctive, are also more or less acquired; they become all the more conscious and intended in proportion as their more complicated nature renders their execution more difficult.

We see how greatly the child is interested in his efforts to stretch out his hands towards a wished-for object, to hold it up, to handle it, to sit down, to stand upright, to walk, etc., etc., from the fact that he never tires of repeating these actions after having once accomplished them, and will even repeat them quite inopportunely " in response to desire in general, however ludicrously insufficient to accomplish the desired end." [1]

The will of a child, being more emotional than intellectual, is generally very little amenable to personal control from the age of three months to a year. At this age desire manifests a more or less impulsive character, the volition of a being without experience, if left to itself, being insensible to the influence of any but the simplest motives. When, at the age of four months, a child has learnt to execute a few special actions with his hands, more than one lesson is needed to form an association in his mind between touching a certain brilliant object and sharp pain, and to counterbalance his strong tendency to touch a bright flame. But at six or seven months, the contact with a prickly brush, experienced several times over in a few minutes, will begin to establish, if only for a day or an hour, one of those associations which, by keeping up the recollection of pain, neutralizes the intensity of desire.

---

[1] Ferrier, *The Brain and its Functions.*

The elements which preponderate in a child's will are impulsiveness and stubbornness ; and can we expect anything else from a little creature who is ignorant of the distant consequences of actions, and who, under all circumstances, obeys only the desire or the aversion of the moment? The following anecdotes plainly illustrate the truth of this.

A little girl of three months old, who had always been accustomed to be rocked to sleep, woke up one day whilst the nurse was out of the room for a quarter of an hour ; when the nurse came back, she found the child in a frenzy of despair : her face was crimson, and her eyes for the first time wet with tears; her screams could be heard fifty yards off. The nurse rushed to the cradle and tried by every means in her power to soothe the terrified child. For some time, however, her efforts were all repulsed, the child's hands pushing her away ; and it was quite ten minutes before her caresses and coaxings had any effect. At last, however, the child seemed to be pacified, and consented to take the breast. But as soon as its appetite was satisfied, its forehead puckered up again, its eyes half closed under its contracted eyebrows, its mouth quivered, and it set off crying again. The nurse then thought the child must be ill ; but turning suddenly round she saw lying at the foot of the bed a heap of plaster which had fallen from the ceiling. She was horrified at the discovery, and naturally settled that this was what had frightened the child and made it cry. She kissed and petted it, and began rocking it to sleep again, and in a very few minutes its cries ceased and it was sleeping peacefully. Then the nurse, who was alone in the house, set to work to clear away the pieces. She had to go down into the yard to throw them away, and a neighbour on the *rez-de-chaussée*, told her that he had heard the noise of the plaster falling (which was very slight) directly after the nurse went out : he had seen her go out, and thought the child was with her. The noise of the falling plaster must have awakened the child with a start, but this could hardly have been the cause of its crying so long : the fact was, she wanted to go to sleep again, and grew more and more distressed at having no one to rock her to sleep, as usual. Had the nurse rightly guessed at the cause, she would have had much less trouble in quieting her. Afterwards, in telling me this incident, she added : "I finished where I ought to have begun ; children are very stubborn, monsieur, about their habits and customs."

All who have had anything to do with the management of babies know well what dreadful scenes often take place during the operation of washing and dressing,—what screams, tears, and struggles. They cannot bear being touched by water, even if it is warm. With some children these scenes begin over again every day, unless by very great skill they can be slipped into their cradles while asleep. I knew one child who, at the age of four or five months, could not be got to bed without the assistance of several people. If one leg was got under the clothes, the other would be kicked out ; one hand had to be held whilst the other was being bandaged up (it was necessary to bandage his hands, to prevent his scratching his face whilst asleep); and the whole proceedings were accompanied by screams, howls, and contortions of the face and body. when he was six months old, his mother, wishing to avoid the almost daily recurrence of these disagreeable scenes, resolved to put him on her bed to sleep for an hour or two after dinner. Instead of this improving matters, however, his fury was all the greater when he was put in his cradle again ; the bed had spoiled him, and he would have nothing to do with the cradle. It ended in their having to let him be on the bed as much as possible, and take care to get him sound asleep before he was put into his cradle at night. If he woke up afterwards, the flickering flame of the night light was enough to prevent his crying and to soothe him to sleep again.

I saw this child again when he was about a year old ; and his nurse then told me that though in five or six matters he was still very obstinate and self-willed, yet he was generally good-humoured and even pliable. If any one smiled at him, he would smile back again ; he made himself at home on anybody's lap ; he had learnt not to *wipe away* certain people's kisses ; and he could be left alone quite happily for hours with neighbours, or even with strangers whom he had never seen before. But this was if his mother or nurse were not present. Directly one or other of them appeared, he would smile at her from a distance, hold out his arms to her, and make efforts to get down if she did not come to take him, or if he were not carried to her. Sometimes they would put him on the floor to go to the person he wanted; and then he would crawl on all-fours, pushing himself along with the help of his knees and stomach. This was a pleasing kind of self-will. In the morning, when his mother had had her breakfast, she used to take him to his father's bed, and there he went through all sorts of little games and tricks, turning over and over, now on his back, now on his stomach, burying his head in the pillow, diving under the sheets, wriggling about like a serpent, mimicking the noise of birds, and giving vent to decided bursts of laughter. This morning entertainment he looked upon as his right. If, when breakfast was over, his mother did not at once take him up to his father's room, even when he did not hear his father's voice, he would call out *Papa !* and then, if no one seemed to understand his gestures and his eloquent looks, there would come mingled screams and sobs and shrieks of *Papa*.

A little girl, now rather a big girl, and of an extremely amiable and sweet disposition, was very difficult to bring up, so her grandmother told me. Up to the age of a year, there was always the greatest trouble to get her into her cradle. It was often necessary to wait a long time for a favourable moment to pop her in, and then the slightest movement would wake her up again ; or if she did not wake up again at this critical moment, she would set off crying like an automaton moved by a spring, the instant she came in contact with the obnoxious cradle. This unconscious crying always roused her up, and she had to be taken out of the cradle again, to avoid the tears and screams which made her mother fear for her health. There was also the same difficulty always over the bath. She would stiffen herself out like a poker, I was told, and scream

like a little fury. Every possible dodge was tried to get
her to stay in ; the nurse pretended to get in with her, her
favourite toys were brought her, etc. ; and when at last she
had been got in, the mother or the nurse must not stir, or
the crying and kicking began again.

I will give one more example at about the same age.
Juliette is twenty-two months old. Her mother has forbidden her to touch the flowers in the window ; she is only
allowed to water them with a child's watering-pot. She
performs this task with a zeal which is only equalled by
her awkwardness. She has to ask her mother's permission
always before doing it; the flowers would be swamped
every day if she watered them whenever she wanted to.
When she has been disobedient, the nurse waters the flowers
instead. Several times a day she is heard saying : "Ittle
dal dood, water fower." The other day her mother was in
the drawing-room with some visitors and the child suddenly
disappeared with her toys. At the end of ten or twelve
minutes she reappeared, her frock and pinafore literally
soaked with water. "Ittle dal dood, water fower," were
her first words on coming into the room. One of the
visitors, on kissing her, remarked the state she was in. Her
mother flew into the next room, where water was flowing
everywhere. This is what had happened : as Juliette had
been disobedient at table, she was forbidden to water the
flowers, and her little watering-pot had been hidden. But
the temptation was greater than her fear of being scolded
for disobeying a second time. She had got hold of her
nurse's watering-pot and turned on the tap to try and fill it
herself; but she had only succeeded in producing an inundation and deluging her frock. She looked so crestfallen
and penitent that her mother's only scolding was to laugh
at her. In the evening she said to her father : "Ittle dal
not be dood, not no longer dood, cause not like water
fower—not like be all wet." The next morning the nurse
gave her back her watering-pot; but she threw it down
saying, "Not dood water-pot, not dood, wet ittle dal, no
more water fower." Her resolution, however, did not hold
out when she saw the silvery shower pouring out of the

nurse's watering-pot. " Ittle dal, velly dood, water fower," said she, and picked up her own little pot.

There is not one of the above-cited examples which does not prove that, whether under the form of automatic desire, or of conscious desire, or of voluntary determination, action in chi̇dren is almost always subordinate to their sensibility. What they want, is what pleases them at the moment, or what they remember to have been pleased with ; what they dislike is whatever displeases them or has displeased them.

In proportion as a child forms its experience, a greater number of sensations, of sentiment and ideas,—that is to say, of determinant motives and impulses,—intervene between the action and the determination, which was before simple and instantaneous. The action, when one is produced, is determined by the strongest motive : actions of this nature are called deliberate, in distinction to those which are impulsive. One often sees a young child hovering between two contrary motives, both of which equally solicit him to action : in general, the hesitation lasts but a short time, especially if both the motives are agreeable. This, no doubt, is for the reason assigned by Delbœuf, " because in this case it is better to act than to wait. Buridanus' donkey, we may be sure, died neither of hunger nor thirst."[1] There is not even in such a case a conflict, properly so called, but merely a concurrence of motives. An almost analogous case is, when a child has to choose between the pain of not having a strong desire gratified, and some compensating pleasure or present which is offered to him. Here regret struggles with an attractive reality ; the child who was crying is on the point of smiling ; then he begins to cry again, and finally he consoles himself and begins to laugh and play. At other times, again, the conflicting elements are more numerous and more distinct; for instance, we find on the one hand these four terms : obedience, privation of a desired object, approbation, caresses ; and on the other hand these four : disobedience, the attraction of the for-

---

[1] See *Revue Philosophique*, Nov. 1881, p. 517.

bidden object, reproaches, punishment. These different motives either evolve themselves separately or are fused together; and under the influence of their representations, impulse is counter-balanced for a longer or shorter time. In cases of this sort the habit of associations formed between certain motives and certain ways of acting greatly facilitates the operation of the will. But deliberating volition, *i.e.*, the reciprocal inhibition of the tendencies to action, tends more frequently in children to revert to impulsiveness, and to impulsiveness dominated by actual and personal suggestions.

When Bernard, fifteen months old, cries without any reason, his father says in a loud voice, " Hold your tongue ; " and sometimes he stops at once. When his father says, *Drink*, or *Walk*, he also generally obeys. But he obeys much more readily when it is a question of anything which amuses him or gives pleasure to others. Thus, when told to imitate the dog or the cat, or a monkey, to clap his hands, to say *No* and shake his head, he does not have to be told twice. If he is forbidden to do anything that pleases him, he obeys less promptly than if it is something that is indifferent to him. He obeys positive orders more readily than negative ones —*i.e.*, when he is told to do something, than when told not to do something ; and this one can easily understand, for in the first case there is no conflict between the will and the order given him. Although he sets himself obstinately against walking alone, I made him take four steps towards me, by holding out to him half a peach which he wanted very much, and which I would not take to him. But his will to do right is as vacillating as his legs. When he had eaten his bit of peach, I made a second trial, and the result was different this time. I offered him another bit of peach and said, " Come to papa, come and fetch the peach," and he came at once, but on all-fours, which is his way of running. He obeys his mother, however, better than his father, gentleness being either better understood by him than sternness, or else having more influence over him. He had let fall a piece of bread, and his father having told him to pick it up, he turned a deaf ear. His mother went up to him, and he wanted to take her hand; but she said to him : " I will give my hand to baby when he has picked up the bread ; first pick up the bread." He picked it up, and then held out his hand to his mother.

Thus we see, even in children less than three years old, a certain faculty, very fluctuating and capricious it is true, "of inhibiting or restraining action, notwithstanding the tendency of feelings or desires to manifest themselves in active motor outbursts." [1] This faculty of restraint, which insures the volitional control of the movements, and, up to a certain point, of the ideas also, belongs to a period of greater maturity, when experience suggests stronger motives to the mind by calling up before it the pleasant or un-pleasant consequences of actions, and when the faculty of attention, as well as the brain centres which minister to it, being more developed, restrains, as by a sort of mechanical inhibition, the sensations, sentiments, and ideas, whose tendency, if not combated, is to produce instant action. We see then that attention is one of the most important elements of volition, which explains the indecision, weakness, changeableness, and caprice of a child's will. [2]

---

[1] Ferrier, *The Brain and its Functions*, p. 232.
[2] I shall consider further on the close relation between the will and the moral sense.

# CHAPTER VIII.

## I.

### ATTENTION.

As attention is the result of an intense or distinct sensation,
and as the organs of young children are not yet able to
prolong these vibratory excitations, shocks of attention
appear to be very rare in babies quite at the beginning of
life. They seem, however, to occur now and then. When I
waved an object at a little distance from the eyes of an
infant seventeen days old, his eyelids trembled and closed,
expressing either fear or the desire to escape from a vaguely
unpleasant sensation. His eyes followed, from right to left
and left to right, a candle that I moved in either of these
directions. The sound of a door being shut, or a loud
voice, made him tremble so much that it was felt by the
person holding him. But the same causes only reproduced
the same effect during the space of three or four minutes ;
afterwards the child was no longer impressed by the colours
of the moving object, the light, or the various sounds ; it
reassumed its habitual rapt expression, its eyes becoming
fixed and immovable as if looking within. After a short
time, all these different impressions, by dint of constant
repetition, become objects of reminiscence, of vague intention, of desire, or of dislike.

The attention paid by children to their sensations and
recollections becomes stronger and easier every day, so
much so that if often appears reflex. When children

seem most to wish to fix their attention, and in fact when
they are most attentive, there is in reality the least need for
their being so. One might well compare an attentive
infant to a kitten, which the sight of some bright object, or
some watched-for prey, will hold for a long time immovable,
its neck stretched out, its paws pressed against the ground,
its body drawn up, its eyes dilated, its upper lip slightly
arched, as if directed to the object of its desire. Here we
have a sensation, or group of sensations, exclusively perceived, repeatedly renewed, and expected; the observing
subject seems to belong less to himself than to the object
he is observing; it is a case of intense reaction, though as
it were passive, of attraction more or less conscious, of the
fascination of an attentive being by the object of his attention. The pleasure which sucking the breast affords a
baby becomes an object of attention. The child delights
in the operation, and as it were listens to and looks at
himself, and feels that he is enjoying himself. This
budding of the conscious faculty, which is designated by
the name of attention, is produced first from the exterior
to the interior; it is an excitation of the nerve cellules under
the influence of an irritating impression; it is not a tension
or an effort from within, but only an adaptation of the
nerve centres to admit a sensation. Attention may be the
result of an action of the will; but its own proper actions
are apart from the will; it is a channel which opens to
external impressions, and which the will may sometimes
keep closed, but which is generally open in spite of it. It
is this muscular and nervous tension, this intense reaction
of vivid impressions, more often forced than voluntary,
which constitutes the primitive character of attention.
Bossuet designates it as "forced attention," and he adds:
" But this is not altogether what we call attention; we only
apply this name when we choose an object to think about
it voluntarily." This only means that attention has degrees,
and that that of which young children are capable, voluntary or involuntary, is always in the beginning induced by
sensibility.

I knew a child who, at the age of a month, certainly

from time to time paid sustained attention to the act of suction. This was evident from the fixed expression of his eyes, which sparkled with pleasure, and at intervals half veiled themselves under the eye-lids. One day his bottle was filled with sugared water; after sucking a few drops he stopped for three seconds, and then began again and went on with the same attentive expression of satisfaction as if the bottle had contained milk. Plain water without sugar did not have the same success; he stopped short after the first mouthful, went back to the bottle after a pause of five or six seconds, and soon turned from it again frowning and pouting most significantly. Here then we see attention applied to the functions of taste.

During the first month, the various movements of the organs, automatic or conscious (I allude only to the pre-hensile organs), are executed in so awkward a manner, they are so vague and undecided, that I have not been able to recognise in them with certainty the influence of attention applied to tactile sensations. It is, however, impossible but that attention should be equally exercised in this manner. In fact, when, at six weeks old, the hands of a child wandered over its mother's breast, face, and hands, the fixed and joyous expression of its eyes, and its wide-open mouth indicated that it enjoyed prolonging these tactile sensations, however vague they might be. At two months and six days the same child would feel its mother's breast and face and keep hold of her finger with evidently voluntary attention. At the same period he began to stretch out his hands towards his mother's breast, when she uncovered it at a distance of two decimeters from his eyes.

I have seen a little girl twenty-eight days and a little boy thirty-five days old, show, by the fixity of their eyes and the movements of the lips at the sight of the feeding-bottle, that they recognised the instrument by which they were fed, and were capable of directing their attention to it. At the same period they used to lift their hands automatically to their face with probably very little consciousness of this involuntary movement; but when the object of the move-

ment was to rub **some part of the face, the** repetition **of it,**
and the accompanying puckering **of the** forehead **and**
dilating **of** the eyes, indicated that their attention **was**
directed to some disagreeable sensations which they **were**
experiencing. A child of one month would look fixedly
for three or four minutes at the flickering reflection of the
light on a table near the window. On the forty-fifth day I
saw him follow with his eyes, after having **first well** looked
at it, a doll **dressed in** light blue which a **little** girl was
dandling **up and down** at **a distance** of **more than a yard.**
Five **days later, the fixity** of his eyes **turned in a certain**
direction indicated that **his attention was concentrated on**
something white, blue, **or violet ; other colours appeared to**
be indifferent to him, perhaps **owing** to temporary Daltonism. When two months old, **he** began to notice red ; but
he still took no notice of black, brown, lilac, or yellow.

This might either have been the result of special **pre-**
disposition to certain sensations of colour, or of **relative**
feebleness in the different organs of sight ! **We must** put
together **the results** of a large number **of** experiments in
order to draw any reliable conclusions. Infants progress
from day to day and sometimes from hour to hour, re-
**vealing, one** after another, faculties which had not at first
been **noticed.** Those faculties which will **one** day be the
strongest are not always the first to appear. I know a case
of a child who did not fix his attention on any coloured
object before the age of two months, but who **at two**
months and a half was as sensible to colour **as are the**
most precocious **children.**

As to the attention **paid to sounds, it shows itself in**
the second fortnight. **Most** babies at **the** age of **twelve,**
thirteen, or fifteen days tremble when they hear **a loud**
noise. "On the 5th of September,—thirteen days **after his**
birth, that is to say,—it was observed **that** Tiedemann's son
took notice of **the** gestures of those who spoke to him (?) ;
their words even called forth or stopped his tears." I have
seen a child sixteen days old sometimes leave off crying
when his mother spoke coaxingly **to** him ; but the rhythmic
movements which she made at the same time had **perhaps**

I

as much to do with pacifying him as her words. One
night, however, she could not succeed in quieting him;
the father got angry and scolded the child, and finally in
a loud voice ordered him to be silent. I cannot say what
was the impression produced on the child, but he instantly
stopped crying.

"At two months and a half," says M. Taine, à propos of a
child whom he had studied, " I noticed a movement which
was evidently acquired. Hearing the voice of her grandmother, she turned her head in the direction whence the
sound came." I have also thought I noticed the same
thing in a child of a month and two days. I was on his
left hand and was speaking in a loud voice; his head
appeared to turn towards the left; and the fixity of his eyes
seemed to express a certain amount of attention to the
sound of my voice. At the age of a month and a half a
little girl-baby would consciously express either pain or the
desire to be fed, by two different kinds of cries. When she
was two months old there was no difficulty in distinguishing
whether her tears meant pain, desire, or anger. Another
child, six weeks old, used to make starts on his mother's
lap at the sound of the violin. When two months old,
the barking of a dog in his room caused him to contract
his eyebrows, purse up his lips, and begin to cry; the dog,
on being petted, changed his barking to a sort of coaxing
yelp; the child listened very attentively, became gradually
calm, and appeared to take pleasure in this new sound.

At a very early age we notice direct correlation between
the strength of attention and the vivacity of the sensations
or feelings experienced. With little children, as with
young animals, those who most easily fix their attention
seem to be those whose nervous excitability is the greatest.
At three months and a half, a very quick little girl, who
could already distinguish some parts of her body, paid
attention to all that went on around her, to noises of every
description, the sound of a voice, a step in the room, the
shutting or opening of a door or window, and to all
colours, even the least vivid, when they were placed within
reach of her sight. A boy six months old, intelligent on

the whole, but lymphatic and not very sensitive, would
hardly look at a nosegay of pale-coloured flowers, placed
rather near him ; I had to wave them close in front of his
eyes in order to arrest his attention.    A very brilliant
flower, however, which I put beside the nosegay, gave him
great pleasure, and he looked at it for one or two minutes.
Presently a cat came on the scene, a kind of animal that he
had not yet seen.    The child uttered several cries of joy,
stretched out his arms, and leant his whole body towards
the animal.    As the cat, however, did not come up to
him, he subsided into an attitude of quietude.    Thus it is
evident that the power of attention is primarily related to
the vivacity of sensation.

The more developed sensibility which produces the
various sentiments, that is to say, ideas, recollections of
sensations, magnified and exaggerated, exercises a considerable influence over the attention of adults.  " With the
greater number of men, the definite direction which intelligence takes, is inspired by sentiment." [1]    What is it to love,
if not to think constantly and with pleasure of some
person or object ?    What is it to hate ?    To think always
with pain of a disagreeable object.    It is passion,—that is to
say, attention constantly excited, or even over-excited, by
feeling,—which makes lovers, artists, heroes, and savants.  I
would modify the saying of Buffon, and I would say that
genius is a long passion.    The little infant, hardly a month
old, is, as we have said before, already capable of feeling
sentiments properly so called, though in a limited degree.
He loves his mother for what he gets from her, if not for
herself ; he loves his feeding-bottle or the breast which
feeds him, the arms which caress or carry him, the faces
which sing to or smile on him, the eyes which speak to him.
He fears, he desires, he suffers, he hopes, he frolics, he
gets irritated ; and these are all so many different excitations of his faculty of attention, already developed by
exercise and by the habit of keen sensations.    But his

---

[1] Dr. Castle, *Spiritualistic Phrenology*, chapter on Education.

moral sensibility, still in a state of vague formation, can only influence his powers of attention in a very slight degree, even when there are the most favourable hereditary predispositions.

This original diversity of the faculties which concur to form or to excite attention, is generally modified by the effect of natural compensations. A very impressionable child experiences too great a number of different sensations for them to be able to transmit durable impressions to the brain, hence a habit of quick and ready attention capriciously or insufficiently bestowed on every object in turn, and thus it is to be feared that the little girl of whom I have spoken above, with the ordinary education which awaits her, that is, very little more than the chance action of circumstances on her own original faculties, will become a mere commonplace or flighty woman, with no definite power of thought, and full of unbalanced and fantastic ideas; superficial, in short, both in heart and mind. On the other hand, the boy described at the same time, less susceptible to ordinary impressions, with sensibility less easily roused, more slow to bestow his attention, and more slow also to take it away, will very likely, under the same educational conditions, develop a clear, firm, and practical mind. He may perhaps have few ideas, but those he has will be firmly held and accurate, because he will have taken his time to form them; and provided the spontaneous development of his mediocre faculties be not checked, he will know well the little he does know. We may add that his organs of attention will have acquired by exercise a power of adaptation which will perhaps compensate for their want of natural quickness. This is a point of great interest to rational educators.

Whatever differences natural dispositions or regular exercise may produce in individual faculties, the general tendency of a child's attention is to be brief and volatile. The concentration of mental activity, the direction of the intellectual vision, look after look, on one subject, is as difficult for a little child as it would be for a valetudinarian to repeat during the space of two minutes the series of muscular efforts which represent the lifting of a heavy weight. What we call an act of attention, is in reality a series of attentive actions, repeated a greater or less number of times during a relatively short interval. We must not, however, exaggerate the astounding rapidity of thought. Its supposed immeasurable velocity has in fact been measured, both in men and animals, by the Dutchman Donders, and others after him. But although the rapidity of thought is not incalculable, it is none the less very great; and to pay attention to any one thing, if only for a few minutes, is to have observed it several times during that short space of time.

Thus, in spite of the services that will, firm desire, and habit render the faculty of attention, by continually increasing its facility and energy, yet even in the most gifted of men it is always liable to fail. Persons the most accustomed to intellectual work are often obliged to force themselves to set to work even at the most familiar tasks. Very few people, I take it, work because work is pleasant to them. It is almost always the urgent motives of necessity, interest, ambition, or duty, which determine them to make and remake the first painful plunges; but once started again in the accustomed groove, the pleasure and excitement of getting on keeps them going and facilitates and sharpens their power of attention up to the moment when lassitude causes it to wane. Grown-up people find a remedy for this periodical weariness and difficulty in intellectual labour, in that law of nature which makes change of work a rest to the mind, just as variety in diet gives a fresh stimulus to the appetite. But a young child, drawn continually in all directions by over-exciting impressions, and endowed with very little power of resistance, has no such antidote for lassitude of the brain. At the smallest effort he is compelled to surrender.

Thus the greatest amount of attention in a child is infinitely little. This faculty of attention, which plays so considerable a part in scholastic life, has hitherto been very imperfectly studied. I know scarcely more than two men, Horace Grant and Edwin Chadwick, who have taken up the subject. Their researches have taught us, that after five

or six minutes in the case of infants, and from thirty to
forty-five in the case of older scholars, the attention becomes
fatigued and intellectual effort fails; that in schools, the
capacity for attention varies with the duration of the classes,
the season of the year, the time of day, the day of the week,
the interval of time between meals and studies, etc. These
observations, however, have done no more than open up the
way.[1] If this question has only begun to be considered
with regard to children in general, it has not even been
thought of in connection with the first period of infancy,
that is to say from birth up to the first attempts at speech.
Long and patient observation will be necessary, to construct
merely the outline of this science of infant psychology, of
which we do not here intend to make an exhaustive treatise
but simply a sketch.

But if children use their attentive powers very feebly,
they use them frequently, and in a rapid but nevertheless
profitable manner. The strong sensibility of their young
brains sometimes makes up for the adults' power of concentration, especially if it be aided by frequent repetition.
"In young children," says M. Luys, "the cerebral cells are
endowed with special histological characters; they are
flabby, greyish, flexible in a manner; they are, moreover,
from the dynamic point of view, virgin to any anterior impression.[2] The sensorial excitation that affects them at
that age must therefore imprint itself upon them more
readily, since it finds them in a state of vacuity, their power
of retention not being put to the test. On the other hand,
in the first years of life the cerebral substance is in perpetual
exercise and organic development. New elements are per-
petually being added to the old ones; and as the new are
most probably derived from their predecessors, we are led
to conclude that the daughter-cells which appear, borrow
from the mother-cells which give them birth an inevitable
bond of relationship, a species of hereditary transmission of
the different states of the mother-cells from whence they

---

[1] Fonsagrives, *The Physical Education of Boys*, p. 176.
[2] This is not altogether our opinion.

spring. . . . It is in these intimate connections between cell and cell, in these mysterious bonds of relationship, that we must look for the secret of the perennial character of certain memories. Thus it is, that certain impressions received in our childhood become the common patrimony of certain families of cells, which maintain them in a state of freshness, incessantly vivifying them by a sort of permanent co-operation. . . . In the young child the impressionability of the cerebral substance is such that it retains, *motu proprio*, all the impressions that assail it, as passively as a sensitized photographic plate that we expose to the light, retains all the images that are reflected on its surface." [1]

It remains for us to examine two questions relative to this faculty of attention, which is said to be the master faculty of the mind. First, Helvetius has maintained the rather Cartesian opinion, that all men are naturally gifted with the attention and intelligence necessary for the same degree of development ; but that attention means labour, and that all men are not susceptible of sufficiently strong passions to change this labour into pleasure. The differences in men thus proceed from inequality of the passions, and not of the faculty of attention. The famous professor Jacolot has taken up the same theme in the present day ; and he seems to have even exaggerated the paradox. According to him, all intelligences are equal; minds only differ in consequence of the different powers of attention, which are in proportion to the will of the individual. The truth is, in a physiological and psychological sense, that we bring with us at birth intellectual and moral orgnizations different in quantity, but not in quality. But it is not perhaps quite reasonable to trust absolutely to first appearances in respect of intellectual and moral inequalities. Is it, in fact, impossible to remedy an incapacity of attention which is perhaps hereditary, perhaps accidental and passing? Can we tell, either, whether these inequalities are not, perhaps chiefly, the result of the training during infancy ? This question, which

[1] Luys, *Le Cerveau et ses Fonctions.* See Eng. Trans. (Internat. Scient. Series), pp. 159, 160.

I shall do no more than briefly suggest here, is a complicated one. To settle it now, either by the laws of heredity or by means of the very meagre knowledge of infant psychology we at present possess, would in my opinion be premature. At any rate I shall content myself with merely suggesting the question here, and shall hope later on to enter into it more fully, taking into consideration all its developments and ramifications.

The second question of importance to those interested in infant education, is to know whether it is possible to predict a child's capacity for attention from phrenological indications. This is a problem connected with the preceding one, and quite as difficult to solve. This much, however, we may say, that without admitting with Gall and Spurzheim a topographically separate localization of the independent faculties, modern phrenology recognises the facts that certain portions of the cerebral hemispheres,—the anterior lobes, for instance,—are always concerned in the carrying on of intellectual operations, and that these operations, identical in nature, evidently show different degrees of complexity in different individuals. Ferrier, who, by his skilful experiments on the brains of monkeys, has thrown so much light on these obscure subjects, believes in the possible localization of the perceptive centres in the hemispheres. With regard to attention, he thinks, like Bain and Wundt, that this faculty implies the activity of the motive faculties. "In calling up an idea" he says, "or when engaged in the attentive consideration of some idea or ideas, we are in reality throwing into action, but in an inhibited or suppressed manner, the movements with which the sensory factors of ideation are associated in organic cohesion. . . . The recall of an idea being thus apparently dependent on excitation of the motor element of its composition, the power of fixing the attention and concentrating consciousness depends, further, on inhibition of movement. During the time we are engaged in attentive ideation, we suppress actual movements, but keep up in a state of greater or less tension the centres of the movement or movements with which the various sensory factors of ideation cohere. . . . In proportion to the development of **the faculty of** attention are **the** intellectual and reflective powers manifested. This is **in** accordance **with** the anatomical development of the frontal **lobes** of **the** brain, and we have various experimental and pathological data for localizing in these the centres **of** inhibition, the physiological substrata of this psychological faculty. . . . The powers of attention **and** concentration of **thought are,** further, small and imperfect in **idiots,** with defective development of the frontal lobes **; and disease** of the **frontal lobes is more** especially characteristic of dementia or **general mental** degradation. . . . The development **of** the frontal lobes is greatest **in men with the** highest **intellectual** powers **;** and, taking **one man with another, the greatest** intellectual power **is characteristic of the one with** greatest frontal development.

"The phrenologists have, I think, good grounds for localizing the reflective faculties in the frontal regions of the brain; and there is nothing inherently improbable in the **view that** frontal development in special regions may be indicative of the power of concentration of thought and intellectual capacity in special **directions."** [1]

**It** is thus beyond doubt that a **marked projection of the frontal region of the cranium** is generally **the sign of natural** power **of attention ; from** which, however, we must not infer, in **the** case **of** proportionate depression of this region, a radical incapacity for attention. There is **no** hereditary defect which education cannot **lessen or do away with.**

## II.

### MEMORY.

Memory is not a purely intellectual **faculty whose** *rôle* **is to** preserve and reproduce the past actions **of its owner. It is** that property inherent in everything, and **especially in** everything that lives, to retain a **trace** of all impressions received. Every impression leaves **a mark, more** or less deep and durable, and capable of **combining in a thousand** ways with

[1] Ferrier, *The Functions of the Brain*, pp. 285–288.

the marks of other impressions in the cellules of the nerve, muscular, and cerebral tissues. Thus, in all parts of the being, there are dispositions, either hereditary or acquired, to remember, that is, to preserve and to reproduce. There is not only a psychological but also an organic memory; sensations, sentiments, affections, states of the viscera, the nerves, and the muscles,—movements and association of movement, all that has once gone on within,—may be revived, whether we are conscious of it or not.

From the moment of its birth, an infant, whose first act of breathing, as we have already said, is pain, who hungers and thirsts, who suffers from the relative cold of the surrounding atmosphere, from the sudden freedom given to its limbs, from the sounds which shock its feeble hearing and the rays of light which strike on its eyes, from the unaccustomed contact with people and things that come near it, and who expresses all these different feelings of annoyance by screams, or rather sharp gaspings, by agitated gestures, and deep flushes over its face and head, is executing automatic actions, *i.e.*, hereditary reflex movements, which have belonged to others before him ; and this is hereditary memory. " Each nerve has a sort of memory of its past life, is trained or not trained, dulled or quickened, as the case may be ; each feature is shaped and characterized, or left loose and meaningless, as may happen ; each hand is marked with its trade and life, subdued to what it works in ;—*if we could but see it*." [1]

All this is found transmitted by heredity, in the actions and gestures of the new-born child, and why should not what is true in respect to apparent movements be also true in regard to other manifestations of human activity, such as sensation, sentiments, and primordial ideas ?

I have often put this question to myself, and not without anxiety, when I have found myself face-to-face with a little child, a mysterious sphinx unconsciously watching me observing it, and whose large calm wondering eyes disconcert

---

[1] Bagehot, *Physics and Politics*, p. 3.

my laborious inductions. I remembered that such and such an action, buried a long time in the storehouse of potential faculties, had suddenly sprung to light, aroused by the fortuitous concurrence of certain favourable circumstances. I have asked myself if I must not set down to the credit of instinct and heredity that which my observations had led me to attribute to consciousness and individual experience. Thus, children have an instinctive faculty for walking, and yet they have to learn the process by long efforts, conscious and progressive. They have an instinct for sucking, and nevertheless, as in the case of dogs and lambs, this operation has also to be learnt by them. Chickens, colts, and calves walk by instinct as soon as they are born, and this is an automatic action, rendered easy by their organization, but which they perfect by exercise and attention.

These reflections have suggested another which has quieted my conscience as an observer, namely (and Darwin is not opposed to this opinion), that the operations due to instinct may easily, at the very outset, awaken a conscious sense ; and that, for instance, the first attempts at sucking or walking may unite in varying proportions reflex influences with conscious efforts. Thus then we may believe unreservedly in the existence of actual activity, even when the gestures and movements seem to be derived from heredity, for they are brought into sub-order, controlled and perfected by the present personality. Thus the little baby who, at the age of five weeks, when his nurse sang him a familiar air or spoke in a caressing voice, began to warble various little sounds, did this by instinct and organic sympathy—the result of hereditary memory, and also by the help of recollection associating together these sounds and the nurse's voice—*i.e.*, the result of individual memory.

It needs only the most elementary powers of observation to see that the extent of the personal acquisitions of a little child, only a few weeks old, is considerable. We have already seen in play, from the age of two and three months, all the various and numerous actions, instinctive, voluntary, intellectual, and moral, which psychologists generally observe at a more advanced age. It is needless to recapitulate these facts, which indicate so evidently how prompt, energetic, and tenacious a faculty memory is at the commencement of life.

It is noticeable also that the recollections most firmly planted in the mind are not always those which exhibit themselves most prominently. The most important, and perhaps the most durable, are those of which custom appears to have dulled the vividness, and whose capacity for revival is dependent on the reflex or unconscious energy of the intelligence. When Georgie stops before the prickly brush without touching it, the idea of a brush is clearly fixed in his mind, and yet nothing betrays the recollection of a painful impression. This is what leads us to believe that nothing is unimportant in the life of man, and that the education of infants should have the same interest for us as that of older children. Most of the ideas which influence the intellectual and moral nature of a man during the whole of his life (as the startling instances of revived memories recorded in cases of hallucination, of hypnotism, or simply of exaggerated passion, testify) spring from the latent capacities of the mind —the mysterious but dominant basis of the soul. Nothing is lost in nature, and, as Mr. Bagehot has eloquently said, each nerve retains perhaps the memory of its past life.

How many of the sudden flashes of sentiment, idea, movement, of happy or deplorable inspiration, which we attribute either to the natural powers of the mind, or to education, or to the effects of example or excitement, are perhaps in truth reminiscences from the cradle. True, I have not been able to establish any instances of memory going back beyond the two first years of life, but that is no reason why none should exist. Language has become incorporated with a great number of ideas anterior to itself; and as, generally speaking, words accompany the reappearance of ideas, we are prone to imagine that the race has had no other ideas than those which are symbolized by speech, and that we have only retained traces of those ideas which were acquired at the epoch and by the means of speech. To affirm this, however, is to deny the existence in animals of a considerable number of ideas which they have no means of expressing. My belief is, that all the ideas which were well conceived and well preserved before the epoch of speech, though they may have been defined and fixed by the use of signs, have been inherent in memory from the most remote epoch.

Such are the familiar ideas of *dog, cat, birds, flowers, milk; heat, cold, roughness, polish, palatable, bitter,* etc., etc. All essential ideas, that is to say, and which, although modified by the posterior work of intelligence, have not the less preserved their date and their rank in the collection of cerebral processes. It is not likely that these impressions, so numerous and so powerful, and which have served as mental exercise during the first two years, should only have been the point of departure of future modifications. These were the first impulsions of a movement which has gone on developing and expanding, and is still going on. Life, in an organized being, does not break off and begin again ; it is a prolonged and continuous series of modifications. Not one of the perceptions of our infancy, whether or no modified by language, is extinct ; they have reappeared, or may reappear, in the intelligence, in the will, or in the sensibility.

In the excellent monograph on memory lately published by M. Th. Ribot I find a kind of confirmation of the hypothesis for which I wanted facts. It is known that the general excitations of the memory seem to depend exclusively on physiological causes, which are often abnormal ones. As to the partial excitations, they too most often result from morbid causes ; but the two following examples would seem to indicate that they can be produced in a sound state of mind :—

"A lady, in the last stage of a chronic disease, was carried from London to a lodging in the country ; there her infant daughter was taken to visit her, and after a short interview carried back to London. The lady died a few days after, and the daughter grew up without any recollection of her mother, till she was of mature age. At this time she happened to be taken into the room in which her mother died, without knowing it to have been so ; she started on entering it, and when a friend who was along with her asked the cause of her agitation, replied, 'I have a distinct impression of having been in this room before, and that a lady, who lay in that corner and seemed very ill, leant over me and wept."[1]

Here is another example :—

"Several years ago, the Rev. S. Hansard, now Rector of Bethnal Green, was doing clerical duty for a time at Hurstmonceaux in Sussex ; and while there he one day went over with a party of friends to Pevensey Castle, which he did not remember to have ever previously visited. As he approached the gateway, he became conscious of a very vivid impression of having seen it before ; and he 'seemed to himself to see' not only the gateway itself, but donkeys beneath the arch, and people on the top of it. His conviction that he must have visited the castle on some former occasion,—although he had neither the slightest remembrance of such a visit, nor any knowledge of having ever been in the neighbourhood previously to his residence at Hurstmonceaux,—made him inquire from his mother if she could throw any light on the matter. She at once informed him that, being in that part of the country when he was about eighteen months old, she had gone over with a large party and had taken him in the pannier of a donkey ; that the elders of the party, having brought lunch with them, had eaten it from the roof of the gateway, where they would have been seen from below, whilst he had been left on the ground with the attendants and donkeys. It may be worth mentioning, that Mr. Hansard has a decidedly artistic temperament."[2]

These facts of normal hypermnesia are still more curious than the much better established facts of hypermnesia from morbid causes. They show clearly, on the one hand, the incalculable power of resuscitation of the impressions received at the beginning of life, and, on the other hand, the close relation which exists between attention and memory, between well seeing and well remembering. And seeing how much sensibility influences attention, it appears evident to us that the quality of the memory depends still more on affectivity than on intelligence. I am led to believe that

---

[1] Abercrombie, *Essay on Intellectual Powers*, p. 120.
[2] Carpenter, *Mental Physiology*, p. 431.

MEMORY.                                127

much that is put down to want of attention is in reality want
of sensibility.   Is it from want of attention or from defective
sensibility that old people, idiots, and young children have
such bizarre and scatter-brained recollections?   Perhaps it
is owing to both these causes.   The feebleness and inter-
mittence with which impressions are transmitted along the
peripheric nerves, and with which they strike on the intellectual, motor, and sensory centres, in all these cases, necessarily cause incomplete and disconnected emotions, ideas,
judgments, volitions, and movements.   However this may
be, the debility of the organs in old people in their dotage,
and young children, shows itself in two opposite ways—dis-
persion and fixedness of ideas and sentiments.   There is a
great deal of both in the period of infancy; and there is no
doubt that the babbling,—at first monotonous but gradually
becoming rather varied,—which babies of two months will
repeat several times a day, after hearing some one speak or
sing near them, and later on their parrot-like chatter, and
also the rambling talk of old people, are the result of some
dominant recollections which pass and repass in their minds,
along with a train of vaguer reminiscences.   I should explain in the same manner the apparently automatic movements of the eyes, arms, and legs, which young children
make with so much animation and so little signification, in
their cradles, in their nurse's arms, or when set upright on
the ground.   These are often impulsions, desires, repulsions,
ideas, sensations, and feelings, recollected more or less consciously, but with alternations of incoherence and importunity, which the purely instinctive need of locomotion does
not sufficiently explain.
    This intermittent importunity of recollections differs from
monomania, in that it is generally connected with objects
that are present, or with the remembrance of familiar objects.
A little girl of eight months, when she has been sucking,
makes a movement with her arm analogous to that of a
person ringing a bell; it is because she is absorbed in the
idea of a little bell which serves her as a plaything ; if she
has it in her hands, she makes it tinkle for a few minutes,
carries it to her mouth, rings it again, and then throws it

carelessly away. When she has shaken her other toys about for a little while she asks for her bell again by the usual gesture, and she will repeat the same manœuvres more than twenty times in half an hour, if she is indulged; this is a purely reflex habit. One day, when I was visiting her parents, I imitated the little girl's movements with a box which was very brilliantly coloured, and made an exciting noise: the child at once wanted the box, and shook it about with evident delight. After two or three minutes I put her little bell back into her hand; she shook it with an air of indifference, and very soon threw it away. I then gave her the box again, and she shook it about for a long time, crowing with delight. This box remained her one interest and occupation for nearly an hour; the little bell was supplanted. Another fixed idea, analogous it is true,.had replaced the habitual one. This temporary possession of the mind by an idea is most noticeable in little children beginning to talk; whether it be that names, by recalling the objects, give them a clearer idea of them and stronger recollection or desire, or that the mere sound of words which have no meaning for them keep their brains and vocal organs on the alert. A relation of mine, when fifteen months old, would keep on incessantly repeating the double sound *a-teau*, which represented to him the idea of boats (*bateaux*), which he had seen on a river, and which had made him wild with delight.

Some months afterwards his vocabulary was augmented by the following jargon, "*Cote noi, toute vilai; cote blan, toute jolie.*" This black *cocotte* and this white *cocotte* were two hens, which he used to go and look at two or three hundred times a day in their yard at the bottom of the garden; the white hen was very gentle, but the black one sometimes pecked his fingers when taking the food he brought them. The boats and the hens were at this time his absorbing preoccupations. Some months later, having made a railway journey, he used to talk every day of the locomotives, the carriages, the whistle, and the steam : *min fer* had supplanted the boats and the *cocottes* in his brain and his conversation. One of the most debated questions of the day, and one

interesting alike to psychologists and pedagogists, is whether
it is possible in early infancy to form an estimate of the intellectual faculties of a child, and especially of its memory
—memory, *i.e.*, in the sense of reproduction of ideas. Gall,
and many others after him, have inclined to the affirmative.
This savant considered that large prominent eyes were the
index of a facile memory (always, be it understood, a
memory for ideas). I believe I have also heard thick eyebrows mentioned as signs of a like facility, and especially
of tenacity of memory, or at least eyebrows which are close
together, and which also denote jealousy. My personal
observations do not warrant me in either adopting or reject-
ing any of these theories.

But I hope that the study of the connection between
physiognomy and transmitted faculties, will furnish the
means of determining, even before children can speak,
whether they have good or bad memories, and that we
shall thus be able to improve the bad memories by exercise
and training. Then, too, it will be possible, and very useful,
to be able from the first months of life to distinguish
between the original aptitudes of the memory and those
acquired under the influence of various circumstances, to
mark the union and the conflict of the one with the other,
in a word, to determine how far exercise and education
modify the memory, as a whole and in its special adaptations.

I must particularly insist on the error of considering
memory as an intellectual faculty having different adaptations, and of believing, for instance, that memory for sentiments is one of these adaptations. The truth is, that there
exists in every human being, not one, but several memories.
Not only do we recollect our past sentiments and affections,
without experiencing them over again, but sentiment only
persists on condition of keeping its place in the memory.
" The most affectionate heart," says Chateaubriand, " would
lose its tenderness, if it did not remember." We can say
that, in regard to sensibility and the emotional life, each one
has its memories, more or less marked and tenacious. In
the same way that each sense, and each intellectual faculty,

K

has its memory, or its associations of memory, more or less energetic according to the individual organisation, so all sentiments and all combinations of sentiment have *their* memories, which are unequally distributed among men. Cæsar, it is said, never forgot injuries : some other people forget neither the good nor the evil which is done to them, and others again only forget the good.

This mode of conceiving of emotional memory, which throws a new light on the genesis of passions and desires, enables us in many cases to penetrate to the bottom of the infant soul. Underlying all the sentiments and tendencies which manifest themselves at given moments of life are the tendencies and sentiments inherited from our ancestors, who have exercised and developed them before us; and the same also may be said of our pains and our pleasures : we know not how much of the past enters into the joys and sorrows of the present. The impressions of our emotional life people our souls and our sensoriums, and are always ready to be revived afresh, whether with or without our consciousness. It is the same with past emotions as with past ideas : consciousness only reveals an infinitesimal part of them. This is what the poet Musset expresses with so much delicacy in the following lines :—

> " Et l'on songe à tout ce qu'on aime
> Sans le savoir."

# CHAPTER IX.

## I.

### ASSOCIATION.

ASSOCIATION is the state of dynamical connection which accompanies the production and reproduction, not only of the facts of intellectual and moral life, but of all the mental, muscular, and nervous facts which are co-ordinated if not altogether brought about in certain fixed regions of the brain. Association is not, properly speaking, a special faculty, like sensibility, emotion, and motor activity. It is one of the apparent or underlying forms of all organic or psychic manifestations. Everything is continuous and bound together, in our perceptions, our affections, our thoughts, and our volitions. We have already seen that all our distinct ideas of objects are equivalent to syntheses or associations of ideas arrived at synthetically. These abstractions and groups of perceptions, which are found in the apparently simplest ideas, explain the nature and conditions of psychic life, which is a series of mental states reproduced and capable of combining with other series of the same kind and with impressions of the moment inserted in the complex woof of former combinations.

It is impossible to know at what moment of the fœtal or intra-uterine life, associations of mental states, properly so-called, begin. Some have certainly been formed long before the time when we first observe them. In fact, what external signs could reveal the numerous impressions which must associate themselves together to produce even a hazy knowledge of the simplest object, such as the feeding bottle, the nurse's face, etc.?

M. Ribot vividly describes all that is involved in a recollection which consciousness presents to us as simple.

"The memory of an apple is necessarily a weakened form of the perception of an apple. What does this perception suppose? A modification of the complex structure of the retina, transmission by the optic nerve through the corpora geniculata and the tubercula quadrigemina to the cerebral ganglia (optic tract?), then through the white substance to the cortex. This supposes the activity of many widely separated elements. But this is by no means all. It is not a question of a simple sensation of colour. We see, or imagine, the apple as a solid object having a spherical form. These conceptions result from the exquisite muscular sensibility of our visual apparatus and from its movements. Now, the movements of the eye are regulated by several nerves—the sympathetic, the oculo-motor, and its branches. Each of these nerves has its own termination, and is connected by a devious course with the outer cerebral layer, where the motor intuitions, according to Maudsley, are formed. We simply indicate outlines. For further details the reader should consult standard works on anatomy and physiology. But we have given an idea of the prodigious number of nervous filaments and distinct communities of cells scattered through the different parts of the cerebro-spinal axis, which serve as a basis for the psychical state known as the memory of an apple, and which the double illusion of consciousness and language leads us to consider as a single fact."[1]

Under these circumstances, then, we cannot hope to "seize in the act" the first associations formed in a child's mind between the different perceptions—visual, muscular, and motive, auditive, or tactile, or between these percep-

[1] Ribot, *Maladies de la Mémoire*. See Eng. Trans. (Internat. Scient. Series), pp. 42, 43.

tions and different sensory and emotional conditions : these facts go on in the mysterious depths of consciousness, if even they do not belong in part to unconscious cerebration. We must be content to point out the most apparent.

Darwin does not notice any positive manifestation of ideas in a child's mind before the age of five months. For instance, speaking of one of his children at this age, he says, " As soon as his hat and cloak had been put on, he became very cross if not taken out at once. At five months his eyes would seek his nurse when he heard her name pronounced." This date, which to me seems tardy, is much nearer the truth, however, than that of ten months, indicated by M. Taine for the appearance of the same phenomenon in his daughter. Tiedemann, rightly or wrongly, notes the beginning of the association of ideas two days after birth. If any one placed his son " on his side in a position to suck, or if he felt a soft hand on his face, he became quiet and felt about for the breast." " Here," writes his father, " association of ideas is evident, the feeling of a particular position or of a soft hand awaking the idea of sucking at the breast. On the 26th of January, when five months old, the growing desire to learn manifested itself more plainly. The nurse, whenever the weather permitted, took him out into the street, which delighted him beyond measure ; and in spite of the cold, he was always very eager for this event. The child soon learnt to notice, that when the nurse took up her cloak it was a signal for going out ; and he smiled and crowed, even in the midst of his tears, every time she performed the operation."

When I put two little kittens (three days old) which I am rearing, on my hand in a standing position, they will remain some seconds without stirring, as if enjoying the heat of my hand ; but soon, the contact of my skin not affording them the same sensation as that of the body of their mother, they begin to shake their heads, show signs of uneasiness, move about on their little tottering paws, and try to suck that portion of my hand which has possibly reminded them of their mother's breast. But as the sucking produces no result, after a few fruitless efforts they become more and more uneasy; they do not feel at home; they have begun to realize the difference between my hand and their mother's warm, soft body, and they begin to pine for their familiar shelter, and to cry out for it loudly.

These few facts, taken at random, show us how very early, both in young children and young animals, the exercise of this physical and intellectual faculty, formerly incorrectly called association of ideas, but which is really the associability of all the actions of the nervous system, either with their congeners or with their neighbours, comes into play.

All the first manifestations of animal sensibility testify to the faculty which the mind has of associating and blending in one the different series of homogeneous impressions which it receives. A baby fifteen days old, who had just gone to sleep, and in whose mouth I put a feeding-bottle filled with plain water, sucked it for a few minutes and then began to make faces, to open its mouth, and finally to cry. The same child, who was lying awake in its mother's arms, was taken up in the same position by its uncle, and began at once to whimper. A child two months old, who already smiled consciously at his nurse, would allow himself to be fed by another nurse, his own being close by. These examples prove that homogeneous sensations were associated in their minds to such a point that they recognised them when they were reproduced, and that, not finding them when they expect them, they suffer in consequence, in spite of the feebleness of their powers of comparison.

Herbert Spencer gives to this remarkable phenomenon the name of integration of the sensations. " A colour, the moment it is perceived, not only irresistibly aggregates with the class of feelings that originate on the outer surface and imply outer stimuli, but also with the sub-class of visual sensations, and cannot be forced into any other sub-class. While being recognised, a sound falls simultaneously into the general assemblage of feelings derived from the senses which hold converse with the external world, and also into the more special assemblage of feelings distinguished as auditory; and no effort will separate it from this special assemblage. And to say a smell cannot be thought of **as a colour or a sound, is** to say that it associates itself indissolubly with previously-experienced smells."[1]

On the other hand, "these simultaneous impressions,—optic, olfactory, acoustic,—received at the same moment, and in several circumscribed localities at the same time, **constitute a series of contemporaneous** *souvenirs*, **which are created and implanted in** the memory; **and henceforth** those vibrations **which were born together, and** were simultaneously conceived, will represent **in the series** of reminiscences, a **definite group,** of which **the** elements, united by **the bonds of a mysterious** federation, **will all** live with the **same life, anastomose one** with another, **and recall one** another as **soon** as one link of the chain is struck."[2]

My two kittens, which the mother had **left to go** and lap up a saucer of milk I had **put down** for her, after tumbling over **and over** for a few minutes, **at last** cuddled together to go to sleep. Suddenly the mother returned and sprang lightly into the box without touching her kittens. Instantly, however, they sat **up, as** if moved by **a** spring; the slight **noise** which the mother **had made** in getting back into **the** box, **and** the movement which she imparted **to it** and which they **had felt,** produced auditory and muscular impressions which became associated in their **brain with** the idea of her presence; and to **the idea of** her **presence** was joined the desire to suck, for **they immediately set to** work at this operation.

When six **weeks old, these same** cats, having made acquaintance with raw meat, precipitated themselves, one **after the** other, on **a** ball of **red** paper, which **they had taken for a** piece of meat, **and were** beginning to eat. The sensation of a particular colour was associated in **their** minds with the idea of food, and with the movements of absorption and prehension, which they already produced by reflex action.

A little child, of four **months and** a half, was in its mother's arms; its nurse, who had **just returned** after a

[1] Herbert Spencer, *Principles of Psychology*, p. 255.
[2] Luys, *Le Cerveau et ses Fonctions.* See Eng. Trans. (Internat. Scient. Series), p. 154.

short holiday, before coming into the room, put her mouth to the key-hole and in her most coaxing voice called little Paul three or four times. The latter first raised his head, then turned to the right and the left, and cast questioning looks at his mother; the nurse went over the comedy again, laughing rather loud. Paul could not stand it any longer, he held out his little arms, and made impatient starts of joy, desire, and vexation; and was on the point of tears, when his mother, to spare him unnecessary teasing, called in the nurse. Thus the sound of his nurse's voice and laugh had become very early associated with the idea of her person, with the pleasure of her presence, of the need of seeing her when she spoke, and with instinctive movements to be taken in her arms.

A little child a year old could not see a hat or bonnet, or any article which appeared to be a covering for the head, without saying, "*Mené, mené*" (*promener*), by which he meant, that some one was to take him by the hand and lead him out for a walk. Whilst playing one day at a table, he took a little round mat and put it on his head, calling out "*Mené, mené.*" He said the same thing also when his aunt touched an umbrella. The word "*peudu*" (*perdu*) is associated in his mind with the idea of any object that he sees fall down, and which he throws away, or which he cannot see when he hears it talked about. One day he took my hand with both his, stroked it, shook it about like one of his playthings, and then, the fancy seizing him to throw it on the ground, he pushed it away and let go of it, saying, "*Peudu*," and looking on the floor to see where it had gone. If a flower is given him to smell, he sniffs it for a few moments, and then immediately offers it to some one else to smell. When he is walking in the garden by the flower-beds, he seizes hold of the stalks of flowers, drags them towards him, and gives whatever remains in his hand to some one to smell. The sight of the smoke which I puff out of my lips when indulging in a cigar, causes him to make a very curious movement of expiration resembling the action of a smoker.

But smoke of all sorts attracts him, and makes him

say "*Mené ;*" and when he is near it, it it is I who am with him, he looks at me with a happy expression, and makes this movement of the lips which his grandfather has taught him, and which I have already noticed in another child.

Thus we find in young children the same kinds of associations as in adults. There is not one of the combinations of associations which have been studied so carefully by psychologists, of which we cannot find at least a faint foreshadowing in a child of six or seven months. We can also discern in little children a sketchy idea of the relations of sequence, which is the principle of the idea of time. A child eight months old, to whom its mother held out her arms while advancing toward it, stopped for a moment, and then held out *his* arms ; she called him, but did not yet go up to him ; he gesticulated with his whole body and began to scream ; she advanced a little nearer and he pushed himself forward as if to hasten the meeting. Thus movements, begun at a certain distance and in his direction, were associated in his mind with the idea of continuation.

A child of eleven months warbles, as a sort of accompaniment, so to say, when his nurse sings a simple little air, which he is very fond of, and which she has sung to him ever since his birth. If the nurse stops in the middle, or does not quite finish the song, the child looks at her with astonishment and stops his warbling ; if the nurse goes on with the song where she left off, the child evinces great delight. These associations of successive sensations, multiplied to infinity, prepare for the formation of the abstract and general idea of time, which develops so slowly in young children, even after they have begun to talk. A child six months old will cry for his bottle as if he were starving, even if he has had it a little while before, as soon as day begins to dawn. The mother says, "He guesses that it is time for his breakfast." No, but the return of light is associated in his mind with the idea of sucking his bottle, and this is why the morning twilight provokes in him the factitious need of a meal. We may notice this kind of association, and consequently a faint rudiment of the concrete ideas of time, in the supposed instinct which causes even very young animals to know the hours of their masters' or their friends' meals. Who has not seen dogs, cats, and birds put in a punctual appearance in gardens and houses, at the hours when meals are going on. It is an indisputable fact that signs analogous to those which give us our ideas of time, that is to say, successions of impressions, indicate to these dumb creatures the opportune moment for returning to their restaurants.

The associations of resemblance are not less clearly expressed by the language of action in little children. A boy of nine months is opposite me. His grandmother has just pronounced the word *papa*. The child smiles at her then at me, and holds out his arms to me. The child certainly did not take me for his father, even at first sight ; but certain general resemblances between me and the father had awakened, at the name of *papa*, the ideas, feelings, and actions which had for a long time been associated with the presence of his father, and which caused the child to smile at me, to hold out his arms, to utter little cries of joy, to embrace me and let himself be fondled by me.

It is by associations of this kind that little children learn very early to see well-known people or things in other people or things, resembling them in some way. In a very remarkable article,—though, unhappily for the study in which we are engaged, the observations are not taken till after the age of three months,—M. Taine notes the following facts, "She sees her grandfather every day, for they show her his portrait drawn in crayon, much smaller than life, but very like him. When any one says to her, "Where is grandpapa ?" she turns towards this portrait and smiles. But before her grandmother's portrait, which is much less like, she shows no sign of recognition.

It is this association of like ideas which suggests to little children those comparisons which we often take for abstract generalizations, and which cause the little chatter-boxes to talk of everything *à propos* of everything, their vivid and inexperienced imagination showing them everywhere the known in the unknown, the like in the unlike. It is this imagination also which makes them afterwards seize so vividly and be so startled at contrasts in objects and actions in which at first sight they thought they saw similarity. It is this which suggests to them analogies of all sorts, which to us seem sometimes so strange and unaccountable, but which are so natural to them, though at the same time they have only an external signification for them. For instance, a little girl who had seen some cockchafers on a tree, asked, "When would the cockchafers begin to bud?" Again, it is the association of analogous sounds which causes children's strong liking for alliteration and consonances; this taste first of all shows itself in meaningless repetitions and later by their delight in music and rhymes.

I do not think it is possible to determine in children who cannot yet talk, the existence of associations based on the law of contrast. Is it not possible that for the fusion of opposite ideas a greater degree of intellectual development and a greater power of comparison is needed than a child of hardly a year old possesses? Children pass from the easy to the easy, from like to like, before advancing from the easy to the more difficult, from like to difference, from analogy to contrast.

Besides these connections, which philosophers have entitled accidental or fortuitous, there are others which they have distinguished by the name of logical or rational, such as the relations of cause to consequence, of means to ends, of signs to the things signified. I wish merely to remind my readers of this order of classification, without attempting to discuss it. I shall confine myself to remarking that, the laws of nature and intelligence being given, there is not one of the different kinds of association enumerated above which does not necessarily exist in every intelligence. The proof of this is, that all or nearly all the forms of association called logical, exhibit themselves, as well as the others, in young infants. What baby is there of three or four months, who, having burnt itself with a candle or the fire, or what child of ten months, who, having had a nasty dose of medicine, will not recoil instinctively at the sight of these objects, which remind them of the pain or disgust they have caused. Here we have the concrete idea of cause associated with the concrete idea of effect. The relation of the means to the end, which in children's minds is confounded with that of cause to effect or succession, is a matter of daily practical application at the earliest age. The sight of their food and toys, of the people and objects who please them and gratify their desires, or the reverse, remind them at every moment that they are the instruments of their pleasure or their pain. When, at the age of three months, they cry for their food, or to be rocked to sleep, they already possess a concrete idea of a means tending to an end ; and what is more, they utilize this knowledge in their own way. They know well enough, long before they can speak, what such a sound or such a modulation of the voice means, what is the signification of this attitude or that gesture ; in a word, the sign and the thing signified are associated in their young minds. As to associations established concerning the relationship of genus to species, we should not be able to find more than very vague rudiments of these in little children whose power of generalization is always rather feeble, needing, as it does, for full play, a fairly large amount of experience combined with considerable progress in the acquisition of speech.

M. L. Ferri, in an interesting article on *The First Three Years of Childhood*,[1] has expressed his opinion that the intelligence of a child exists in a degree and is exercised in a sphere proportioned to the degree of experience of which the child is capable before it has acquired any abstract notions ; and that human intelligence, when it is only founded on the sensitive associations common alike to men and animals, consists essentially in an intuitive activity, which distinguishes and combines, affirms and denies the connection either of sensations or of things, and is not limited to a passive reception of corresponding associations. This is also my opinion. I think it is the natural organization only which predisposes a child to modify certain actual ex. periences according to the dynamic impulsions which an.

---

[1] In the *Filosofia delle Scuole Italiane*, Oct. 1879. See in *Revue Philos.*, April 1880, the review of this article.

terior experiences have deposited in its organs under the title
of latent energy. This would still only indicate a greater or
less facility of accommodation in the young being to new
experiences, and perhaps the necessity for him of certain
associations which are imposed on him by his specific
organization, as much as by the nature of things. At any
rate, I agree with M. Taine that the associations of ideas
formed in the mind of a child of ten months do not go
beyond the scope of animal intelligence ; but this may be
because I grant more to the latter than does M. Taine, or
even Darwin, at any rate in the above-mentioned treatise.
Darwin sees a marked difference between the aptitude of a
little child and that of the most intelligent adult animal, in
forming associations due to instruction and associations
spontaneously produced. I cannot subscribe to this. Does
the dog who runs from the bottom of the garden on hearing
the word *sugar,* show a faculty of association inferior to
that of a child of six or seven months, who moves his head
from right to left when one says to him, " Shake your head "?
I think not.

I have had cited to me an example of association furnished
by a little French girl in Vienna. She was two and a half
years old, when she heard related a detailed account of the
assassination of the Czar Alexander II. All the events of
the narrative were firmly imprinted on her mind ; but she
confused the assassins of the Emperor with the men who
swept away the snow on the scene of the assassination. A
fortnight later, snow was falling in Vienna ; the window
being open, the child looked out in the street and saw
several men occupied in sweeping away the snow. Suddenly
she cried out, " Oh, wicked men, they have killed the
Emperor ! me like to see poor Emperor buried !" But
why should we refuse to believe that animals which cannot
speak are not susceptible of analogous associations ? Here
is an example showing how durable connected recollections
may be in certain animals. "A turtle-dove, which had been
very carefully brought up by a lady from London, was given
to an officer who carried it away with him on an expedition.
The bird remained absent a year and a half. When it was

brought back after this interval of time, it heard its former mistress speak before it saw her. At the first sound of her voice it showed signs of extraordinary impatience, and fluttered about the cage, a prey to excessive emotions." [1]

Philosophers have often remarked the influence of association of ideas on sentiment. Thus, the word *mamma* will make a child smile, whereas that of *papa* will only make him turn round quickly, to see the person whom he loves with more of respect than tenderness. In general, the habits founded on associations which are the most frequently repeated appear to be the strongest. But there are cases where the vivacity of the emotions engenders associations suddenly, and, in consequence, establishes lasting habits. Locke speaks of a little girl whom her mother was obliged to whip, in order to cure her of the bad habits she had learnt from her nurse. A single day's correction sufficed. Painful affections, as we can easily understand, have an influence proportioned to their vivacity, on our moral habits especially. Hence those who have the bringing up of children should keep them as much as possible out of the way of terrible spectacles and alarming tales.

In order to well understand the mechanism of thought in children, and to be able to interpret at every turn their gestures, words, and actions, we must not overlook that law of association which causes memory to touch over and over again on the culminating points of a series,—an abbreviative process which may be succeeded by the detailed analysis of intermediate and secondary points. This tendency of the mind, which enables the adult to construct artificial chains of ideas, manifests itself spontaneously in children, and it is possible to follow the formation of the chain up to a certain point. I will cite one among a thousand examples. A lady had gone to call on a friend and had taken with her her little boy two years and four months old. After the child had well inspected every corner of the drawing-room, he slipped through an open door into a little passage which led into an adjoining room. The door

---

[1] Frank Buckland, *Curiosities of Natural History*, vol. i., p. 183.

of this room being half-open, he went a few steps into it, and in so doing upset a chair. This noise awoke the husband of the lady, who, being either tired or indisposed, had lain down on his bed in his clothes with his nightcap on his head. "Who is there?" cried the sleeper, disturbed in his slumbers. The sudden sight of this bizarre apparition caused the child inexpressible fear and surprise. He ran out of the room, took refuge with his mother, and pulled her dress repeatedly, as if to make her come away. Soon after the mother went home with the child; and the whole way along the child talked of nothing but *Pierrot* whom he had seen lying down, and who had frightened him very much. More than three months later, when this lady came to see the child's mother, the child asked her abruptly, "How is Pierrot? Is he sleeping on his bed?" The lady answered that he slept in his cage on a perch, and not in a bed. "Is Pierrot naughty? he frightened me very much." The lady thought he was speaking of her sparrow. Another time, she answered him that Pierrot was dead. "Then," said the child, "he is no longer in bed." The lady still thought he meant the bird, which had just died. The child's mother burst out laughing, and explained the misunderstanding. These were the salient points of that association of images and ideas so alarming to the child. To all these prominent recollections there belonged a number of others, which came back capriciously to the child when he talked of this redoubtable Pierrot.

Another question of importance is, whether it is easier or more natural to recall a series of incidents in the same order in which they occurred or in the inverse order? In my opinion, this last kind of association is essentially artificial; but it is not equally easy, or at any rate habitual, to different orders of mind, apart from all inequality of intelligence. One of my friends, a scientific man of superior mind, sent me the following reflections on this subject: "I remember that Taine asserts that memory can reproduce ideas and incidents in the inverse order to that in which they first arose. He cites as an example a journey along a particular road, all the events of which can be recalled by the memory in an order exactly the reverse to that in which they happened, that is to say, beginning at the point of arrival and going back to the point of departure. A manifest error. I have vainly made the experiment, and I have realized that in a series of objects or impressions, *a*, *b*, *c*, *d*, *e*, *f*, if I wanted to go backwards from *f* to the preceding object, I was obliged by a rapid flight to imagine myself as having reached *e* or even *d* before getting to *f*. To pass from *e* to *d*, I must first recall *c* or even *b*. Thus, having learnt by heart the series of stations from Bordeaux to Morceux, it required a tremendous effort to recall the inverted series from Morceux to Bordeaux, and I was only able to construct it by the help of fragments of the direct series." Such a conclusion surprised me in a man of science, and one thoroughly initiated in the methods of natural science. Must we conclude that he was wanting in capacity to accomplish the feat, or was it only want of habit? It may be that certain minds are naturally more apt at it than others. As far as concerns my friend, I verified by experience that this operation is as difficult for him as it is easy to myself. I made the experiment *à propos* of a road which was well known to us both since our childhood ; taking a distance of two leagues from a town to a village, I repeated, both inversely and otherwise, every little road on the left hand and on the right, which diverged from the high road. My friend made some mistakes even when beginning at the point of departure, but he would have made a still greater number had he attempted the reverse way. This faculty, which I am convinced is essentially an acquired one, but which is also more or less a natural gift, must be very feeble in children. But that it is entirely left out at the beginning of life, nothing seems to show.

It is association which makes the unity of our mental existence, by establishing a natural bond between all the various parts of which it is constituted ; and it is to association that we must look for the formation of the habits, judgment, character, and morality of children. Association presents us with an easy means of exercising the memory methodically, of verifying its acquisitions and its aptitudes, of facilitating its play, and rectifying its errors, by the facility which it affords a child's mind—especially when influenced by an adult will—of ascending or of descending the chain of its ideas and sentiments. Do what we will, however, we shall never arrive at knowing and being able to guide more than a very limited number of the associations which work in a child's mind. The important matter is, to know that we can discover and establish a great many essential ones, those which are the most apparent and habitual, and that we can thus to a certain extent move according to our own will the secret springs of the young character.

We must not give in to Fénélon's opinion, that it is enough to awaken the curiosity of a child and to heap up in his memory a mass of good materials, which will combine of themselves in due time and which the brain when more highly developed will arrange in systematic groups ; we should endeavour as far as possible to control the first impressions which sink unconsciously into a child's mind, but still more careful should we be in the selection of those later ones which we try to inculcate on him, and of the links which we wish to establish between such and such perceptions, sentiments, or actions ; for the older a child grows the less must we count on the innate tendency of just ideas, suitable sentiments, and useful impulses to combine themselves in logical and durable associations. To a child's mind every combination is logical and moral from the mere fact of its existing. From infancy therefore we must keep careful guard over the formation of those associations over which we have any hold. Why is it that domestic animals, —so far as they have escaped man's training and follow the natural impulses of their organization combined with the direct action of external objects,—appear to us often to show proofs of a surer and quicker judgment in the things which are useful for them to know than most human beings would in like circumstances? It is because the instincts in each species, and their variable development in each individual, are subject less to imitation and to education than to personal experience, and to the latter less than to hereditary experience ; and by experience we mean associations firmly

L

combined and in a very limited space. Their logic is more limited than ours, it is circumscribed within a small sphere of relations, but in this sphere it moves easily ; it is founded on the natural relations of things, on associations truly experimental.

As much may be said, in many respects, of savages. It is not only by the acuteness and special adaptations of their senses that they are superior to us, but by the judgments, limited it is true, but founded on practical experience, which they bring to bear on the impressions of their senses. Not only are some of their senses,—sight and hearing for instance,—of much wider scope than ours, but their daily life places them in circumstances fitted for the exercise and development of their sensorial judgments. Concerning the Arawaks, Hillhouse says, "Where a European can discover no indication whatever, an Indian will point out the footsteps of any number of negroes, and will state the precise day on which they have passed ; and if on the same day, he will state the hour." . . . "Along with this acuteness of perception there naturally goes a high degree of skill in those simple actions depending on the immediate guidance of perception. . . ."[1] Such is the result of frequent or minute observation of a limited number of objects. We must therefore be careful to select experiences for children who cannot choose for themselves, and this not only with a view of developing little by little the sensorial faculties of our species till they equal those of savages and certain animals, but also of developing, side by side with the perceptions, the faculty of judging and acting rightly.

To give a child very little to observe at a time, but to make it observe that little well and rightly, is the true way of forming and storing its mind.

---

[1] H. Spencer, *Principles of Sociology*, p. 88.

## II.

### IMAGINATION.

Representative or reproductive imagination, or the return of vivid impressions, images, forms, **sounds, colours, objects,** persons and places which have keenly affected us, begins to work very early in life. The child, hardly a month old, who recognises his mother's breast at a very short distance, shows, **by** the strong desire it has to get to it, that this sight has **made an** impression **on** it, and that this image **must** be deeply engraven on its memory. The child who, **at the** age of three months, turns sharply round on bearing **a bird** sing, **or on** hearing the name *coco* pronounced, and looks **about for the** bird-cage, has formed **a** very vivid idea of the bird **and the** cage. **When, a** little later, on seeing his nurse take her cloak, **or** his mother wave her umbrella, he shows signs of joy and pictures to himself a walk out of doors, he is again performing a feat of imagination. In like manner, when, at the age of seven or eight months old, having been deceived by receiving a piece of bread instead of cake, on finding out the trick, he throws the bread away angrily, we feel sure that the image of the cake must be very clearly imprinted on his mind. Finally, when he begins to babble **the** word *papa*, at the sight of any man whatever, it must **be** that the general characteristics which **make** up what he calls *papa* are well fixed in his imagination.

The terrors which young children and animals **are subject** to, and which are **as** vague and unaccountable as they are strong, are in both cases sure indices of the workings of the imagination. How can we otherwise explain a child's dreams, the sudden tremblings, the screams, the sobs, the **smiles, the movements to seize hold** of an object **or** to repel it violently, **which we observe in** an infant of three months while asleep, **and** which resemble **the** actions produced during its waking hours by fear, pain, hunger, desire, and joy? It may be supposed that these recollections of objects with which he has already **come** into contact came to him also during the diurnal exercise of his faculties, but more as means for identifying analogies and various feelings, than as distinct images in his mind. The absorbing preoccupations of their short waking moments do not leave children as much time as adults for imagining recollections or for waking dreams. Thus, a little baby of two months and a half, who had been held in his father's arms for ten minutes, stretched out its arms eagerly and with evident delight towards its mother when she advanced towards it. The numerous impressions with which the father had occupied and distracted it during its mother's absence, had they not prevented it from thinking of the absent one and imagining her near at hand? Happily for children's peace of mind and delicate health, they are rarely left to themselves while awake.

From my sitting-room I have often heard the children of working-people, whose mothers had been obliged to leave them alone for a couple of hours, screaming distractedly in their cradles; their agonized shrieks of *Mamma, mamma,* as soon as they can utter the sound, indicate clearly what was the nature of the images and sentiments which distressed them several months before, when, on finding themselves in the same isolated state, they had no power of expressing their grief and anger. If children exercise their imagination when awake, everything leads us to believe that they exercise it still more during their sleep. This is in my mind the most favourable time for the intellectual labour of little children, labour which must be frequent and peaceful, and whose value is almost always in direct ratio to the apparent repose of the exterior organs. It is then that those hallucinatory sensations without any present or definite object, those vivid reminiscences, those countless associations— flashes of abstraction and generalization—those collocations verging on comparison, those objective judgments and reasonings, resuscitate, under the more or less open eye of consciousness, in the fibres of the brain, charged as they are with recent impressions and along which the blood ebbs and flows rapidly, awakening incessantly the vitality which is scarcely affected by the paralysis of slumber.

Let us study the transition from reproductive to productive imagination, which is also called poetic or creative. As soon as ideas no longer present themselves in the order in which the intelligence has first perceived them, as soon as something has become altered—were it only one link suppressed in a series of associated images—there may be said to have taken place an artificial mental composition, a modified conception of reality, a spontaneous work of productive imagination. These free combinations of images arise spontaneously in young children ; and is it not often the same with adults, who are visited by so many unexpected inspirations ? The little Tiedemann, when five months, old "guessed," his father says, when he was going to be taken out, and showed signs of delight. His biographer sees no more in this than an association of ideas ; but there is more in it. Mixed up with the act of judgment which the word *guess* presupposes, there is a very vivid conception of some of the recollections attached to the idea of going out, which produces in the child's mind a series of delightful pictures. Now, supposing the child to be awake, this conception would probably remain for some time in exact correspondence with the reality. But during sleep the capricious mobility of the ideas would introduce into these images of reality some circumstances and details of -a fantastic description. Certain details would be added, others would disappear; time, places, objects, would be transformed in a thousand unforeseen ways. The impressions of the reality would become, to a sleeping child, a fictitious reality.

Let us picture to ourselves the working of infant intelligence during dreams, inferring, as analogy will allow up to a certain point, from a child who can speak to one who is still mute.

The infant has just slept for a quarter of an hour : this first slumber, consecrated to the repose of the organs, has been very profound ; a flow of blood to the brain, occasioned by some internal trouble, half wakes him up ; the nurse rocks him, and in two or three minutes he is again asleep. As the first sleep has already in some measure repaired the losses of organic life, this second sleep is lighter and more favourable to dreams. Vague sensations, transmitted from the periphery to the brain, excite a revival of ideas, and the play of associating and dissociating images has already begun. A fly settling on the child's face, a movement of its bed curtains, a sensation of contact caused by the sheets or its nightgown, a ray of light falling on its closed eyelids, a noise in the house or in the street, the muscular sensations caused by one of the child's automatic movements, an impression surging up from the depths of the organs of vegetative life—in short, the slightest circumstance foreign to the ordinary life of the brain, will suffice to awaken its functional energy, and set the child off dreaming. Instantly memory recalls to him any of the most striking incidents of the day—the green bench, for instance, on which his nurse sat with him; and with this comes a whole train of associated images. A large tree with waving branches, a beautiful white cloud on a bit of blue sky, the smiling face of a child who kissed him, let him handle his toys, and play with him, then a dog, spotted with white and brown, which came and put its paws on the child's frock and licked his face; then a man in a red and blue uniform, whose large sword glittered and made a great noise, then another man like the first, who passed very near the path, sounding a drum; then a horse galloping the same way, then some more men in red and blue with gleaming bayonets on their shoulders, then a heavy cart which bumped on the pavement with a loud noise, and directly after that an ugly woman with a smiling face, offering cakes for sale out of a basket. We are supposing that all these salient events, which happened whilst the child was out of doors, have come back to his mind in the order of succession of the real impressions. But the reproduction of these recollections—the illusion of sleep assisting—has made of them a simultaneous picture. This in itself would be a notable modification, were there no other, for imagination to have introduced into recollections of real incidents. To modify still further this chain of associations, it needs only a fresh excitation, coming either from outside or rising from within, which shall combine in the brain old recollections with new ones, separate what was united and unite what was separated, and form disproportioned and incongruous associations of ideas ; and the reproductive imagination will have given place to the productive imagination.

If the series of impressions conceived in the dream were a faithful imitation of the real impressions, why should there be those violent contractions of the face, those contortions of the limbs, that wild laughter and those piercing screams and convulsive tremblings so painful to witness, and all those evident signs of intense sensations and emotions which the realities, the recollection of which forms the tissue of the dreams, did not excite in the child? A considerable change must therefore have been produced in his hallucinated brain, a change affecting the proportions of the images and their mutual relations.

For example, the large horse will have taken the place of the dog; with awful neighings, he advances his gigantic nostrils close to the child's face. The cake, a piece of which was snapped up by the dog, will be seized by the huge mouth of the horse, which will then gallop off to the cart carrying away the nurse and the little child ; and so on through the whole chapter, until the horror reaches a climax, and the child awakes with a loud scream.

But his half-open eyes have perceived, transfigured as in the dream, the gentle face of his nurse, who is rocking him and whispering comforting words in his ear. The charm takes effect, and the child goes to sleep again, and resumes the thread of his dream interwoven now with joy and happiness ; the green bench reappears, and the little children, dressed in white and blue and pink, with bright eyes and rosy cheeks; cake and toys are given to him, and so on, and so on, at the will of the capricious fairy, who is the association of ideas. Such is the rudimentary working of creative imagination.

Dreams are the poems of children, who, even when awake, are always more or less of poets. Ah! the strange dramas, the merry comedies, the sparkling idylls, the dismal elegies, the thrilling odes which have haunted in their cradles the brains of future poets! Fictitious inventions which will perhaps return to them later in life without being recognised, when they think they are imagining quite new and original combinations. For the adult, as well as for the young child, what we call creative imagination consists in separating, combining, cutting down, amplifying, abridging, exaggerating, and juxtaposing in a thousand different ways, and often in an order the least intended, former perceptions, judgments, and reasonings, in order to create out of these materials an inner world quite different from the outer one, though made after its image.

In the waking state the exercise of this faculty shows itself by manifestations of diverse kinds.

## III.

### SPECIAL IMAGINATION.

Every sense has its special imagination; but psychologists tell us that the senses of sight and hearing call up more vivid and distinct images than the other senses. "This is because sight and hearing play a more important part in our existence than taste and smell; incessantly employed for the most various needs, these senses are seldom left inactive. Besides which, the nerves set in vibration by auditory and visual impressions are finer, more distinct, and more numerous, and their ramifications in the cerebral substance are more systematic." [1] Moreover, it is the images perceived by these senses which dominate in animal intelligences, and in human intelligence in a state of rest, or of relative suspension of thought, such as reverie, natural sleep, and hypnotism. In induced somnambulism, as M. Richet has said, "the medulla acts when the brain has become powerless;" and in addition to automatism of movements, one of the most characteristic phenomena of this state is the seeing of images. A child in its ordinary state is what a somnambulist is by accident; in the one and the other the superior nerve centres are but little active and balanced. It is not astonishing that a child's imagination should respond

---

[1] H. Jolly, *Cours de Philosophie*, p. 136.

to external excitations, and *à fortiori* to those of sight, which are the most vivid and numerous. The habitual state of little children is that of waking dreams.

Mr. Galton has made some interesting researches on the mental representation of visual images. The result has been to prove the existence of natural varieties of mental disposition in different races, sexes, and ages. It appears to be established, on the testimony of learned men themselves, that to the great majority of them mental imagery was unknown, and that they looked on him as fanciful and fantastic in supposing that the words 'mental imagery' had more than a metaphysical meaning. On the other hand, people of the world, women and children, on being interrogated by Mr. Galton, "declared that they habitually saw mental imagery, and that it was distinct and full of colour." It has also been established, that colour is more easily imaged than form, especially with children, but that the faculty of imaging colour disappears sooner than the other. These are facts of the highest importance in the training of infant faculties and intelligence, if we allow that imagination is always associated with the other phenomena of psychic life, and that in children it dominates over these others and carries them away with it.

Imagination can also reproduce, and sometimes with great power and vividness, the perceptions of sound. The tone of a human voice, the noise of an animal, the cracking of a whip or the rumbling of a carriage, the ringing of bells, the loud chords of an orchestra, haunt us still when the sounds themselves are silent. We can recall the airs that we have heard played or sung; when we hum the scale, we have every note of it in our ears. A musician composes in his mind, and when he tunes his instrument, he compares its sounds with those he hears in imagination. We also hear mentally the words which express our thoughts, so that it has become almost impossible to us to think without words, a phenomenon designated by philosophers under the name of *internal speech*. The auditive imagination, more or less developed, according to the infant's organization, plays a very important part in its intellectual evolution, and a very important one also in its emotional, moral, and æsthetic evolution. We shall have occasion to refer to this later.

All recollections imply mental images more or less distinct and vivid ; and we could not remember or recognise a single taste or smell if the senses of taste and smell had not also their special imaginations. There is no doubt that we represent flavours and odours to ourselves in a particular manner. If a *gourmet* can taste and smell in imagination, a *gourmand* can enjoy his dinner in advance. The recollection of sensations of smell, although generally feeble, may sometimes be vivid enough to make us take them for reality. People with a very delicate sense of smell can enjoy in anticipation the scent of a rose which they see, or that of a bunch of violets which they are about to buy. We have shown above that this kind of imagination is already intellectually and emotionally active in some young children.

There is also certainly a tactile imagination, a muscular imagination, and a thermal imagination. The imagination of touch is exemplified in people born blind, who can picture to themselves, without the help of sight, the objects around them, who read raised letters by means of their fingers, who possess a tactile system of geometry, and represent to themselves tangible figures as we do visible ones.[1] These three kinds of imagination, which superficial observers confound in one only—tactile imagination—are powerfully exercised from the first days of life. We have already devoted a few observations to them in the chapter on emotional and sensual perceptions. We will content ourselves here with reminding parents and educators of children of their extreme susceptibility to the different impressions in question. A child of three months will already distinguish people from the manner in which they touch, stroke, hold, or carry him. At six months he already appreciates very imaginatively the little punishments that are inflicted on him, the people in connection with these punishments, and the objects which are the instruments of them. There are some

---

[1] Diderot, *Lettres sur les Aveugles.*

children whom the sight of a doctor who has attended them will drive almost mad. If there are some children who are very little sensible,—one might almost say insensible,—to blows, it is quite the contrary with the greater number. I can never help laughing when I see the peasant women in my own country frightening their little pickles of two or three years old, or more, with nothing but these words : " *Qu'où té péli*," or " *Quet ba cosé*" (crude Gascon formulæ, not easy to translate). Whoever is inclined to doubt the powerful effects of tactile imagination, has only to read the retrospective memoirs of a child, now almost an old man, who has not forgiven his mother the " thrashings " of old days.[1] It is a happy thing, moreover, that tactile and muscular impressions do leave deep traces on the child's imagination. If this is the secret of some of his little troubles and sorrows, it is also the cause of his greatest pleasures. We may often notice in little children between fifteen months and four years old an inexplicable kind of hyper-excitement of the tactilo-muscular imagination ; they feel a need to cuddle up against anything or anybody, inanimate objects, people, animals. " Let us squeeze each other," said a child of two years old to a friend of its mother's, proceeding at the same time to hug her tightly. Is this play, or a superabundance of vitality, or unregulated and unconscious impulsions of affectivity ?

The influence of sentiment on imagination and of imagination on sentiment are facts so well known, and which are so frequently the cause of affectionate *rapprochements* between grown up people and children, that they need not be dwelt on here in connection with children. The affections, sentiments, passions, desires, fears, aversions, pleasures, and loves of children have a continual tendency to embody themselves in people and animals and inanimate objects ; they appear before their eyes as images in relief, varied, capricious, and incoherent, nearly always false and exaggerated, often terrible, sometimes enchanting.

---

[1] Jacques Vingtras, *L'Enfant.*

At the age of a year, when a child is not yet a "speaking intelligence," and even after he has begun to speak, *things* present themselves to his mind more than the signs, terms, or gestures which express them. Everything is reproduced in his mind in pictures; he thinks in images, he measures the truth of things by the vivacity of his impressions. Owing to their power of scenic imagination, which brings before their mind's eyes the salient features of real life, and awakens in their hearts the corresponding sentiments, little children who can speak, and doubtless also those who cannot, have frequent moments of retrospection, which affect in a greater or less degree their happiness and their moral health, and consequently, also their physical health.

A little girl, two-and-a-half years old was heard relating with all the seriousness of a grown person, with touching sadness and with tears in her eyes, the acts of brutality which her father sometimes committed towards her mother. "Naughty papa," she said, "very naughty papa! He always does like that to mother; he throws her down, and then I cry. He is very naughty!"

A little boy, between three and four, saw some fir-cones in a park. "Pretty fir-cones!" cried he; "there are some like that at Arcachon, at the sea-side. I went last year to Arcachon with papa and mamma. I played a great deal, and was not much scolded."

What has struck all observers of little children is, that with their aptitude in recalling, even with emotion, all that they have seen and heard, they are unable to localize their recollections in time or space. These Kantian forms of understanding are very incomplete categories at this epoch of rudimentary analysis, abstraction, and comparison. A child of two years, who passed nearly every day along the same street on the way to the *Jardin des Tuileries*, was one day taken by his nurse to the garden of the *Luxembourg* instead. The instant he caught sight of the garden railings he cried out, "*Châdin Tiri.*" Although young children are chiefly engrossed with what they are doing at the present moment, they have also a vague idea of the recent past as such; but this idea must have some actuality for them, in

order to interest them. **They will say,** "I did that yesterday," when it was done in **the** morning. The future which
is not very near, or which does not appear so to them, has
no influence over their imagination. A child of **two** years
said to his mother, "**As** we are going to Royan to-morrow
to see grandmamma, why don't you dress at once?" But
at three years old, this same child, **who** was very happy in
a temporary residence, and **who** knew that his parents were
going **to leave it in** a week, tried **to** put back the day of
departure **by an illusion of** his imagination, which was concerned in seeing that day deferred as long as possible. He
used to say to his aunt, "We are going to Bordeaux in
eight days, **not to-morrow, no, but in a** very long while."
Let it then be our endeavour **to make** this present, which
is everything **to** children, happy and joyous; **let us** by all
means **insure to** them that happiness which costs so little,
and which is the first condition of their moral progress.
Happy children are good children.

Rousseau has written some beautiful passages **on the**
happiness which is due to children, **who** may perhaps never
**know it later.**

**One of the** most terrible diseases that mankind is subject
to is insanity, **and facts** authorize us in thinking that the
youngest child is **not** more proof against this form of
disease **than** against any other. All that we can say **is,** that
it is relatively less frequent in infancy. M. Compayré has
collected from writers on **lunacy a** number of observations
to which **we call the** attention **of our readers. Convulsions**
may, **up to** a certain point, **be** considered as the first stage
of insanity in children, whether arising **from constitutional**
or accidental causes. They have all the **external character-**
istics, and often all the intellectual **and moral** consequences
of mental disease. Trousseau **cites a curious** example.
" A doctor, called in to attend a young child in convulsions,
advised that its cap should be taken off. He then saw a
piece of thread lying on the child's head ; and in trying to
remove it he dragged out a long needle which **had been**
deeply buried in the cranium. The **instant the needle was**
taken out, the **convulsions stopped.** Convulsions, however,

more often arise from constitutional causes, from cerebral affections transmitted by heredity. They appear in individuals who are endowed with special nervous susceptibility, which is transmitted by birth from the forefathers to the children, and which manifests itself now by one phenomenon, now by another—by convulsions in children, by epilepsy or hysteria in adults. On careful investigation, it will perhaps be found that there is not a single family where madness exists in which convulsions have not played a certain part. Even in families which are only highly nervous, and who have yet had no cases of madness in their midst, the appearance of infantine *eclampsia* manifesting itself successively in several children, should be considered a very bad symptom." It is, as it were, the first signal of the possible invasion of insanity in a family hitherto sane.[1]

Hallucination, which is the second degree of insanity, rarely shows itself in children by unmistakable signs. If it does take possession of a child of three or six months, it may easily be confounded with the disorders consequent on over-excited sentiment. We can nevertheless point to cases where the absorption of poisonous substances has thrown young children into nervous states analogous to the state of hallucination. Dr. Thoré is our authority for the following case :—

"A little girl, fifteen months old, had swallowed, in her mother's absence, a considerable number of grains of *Datura Stramonium.* Almost instantly the child was thrown into a state of agitation which frightened her mother very much. The doctor who was called made the following depositions :—'A great change has supervened in the visual organs ; the child seems to be deprived of sight ; she does not look at any of the objects around her, and pays no attention to things which used to please her and which she was in the habit of asking for. A watch is shown to her, and some of her toys ; but they do not attract her attention, while, on the contrary, she appears to be reaching after

---

[1] Compayré, *La Folie chez l'Enfant. Revue Phil.*, Dec. 1880, in the *Annales Médicales.*

imaginary **objects at some distance from** her, and which **she** tries to get hold of by constantly stretching out her **arms** and clutching with her hands. She **even** raises herself **up** by leaning on the sides of her cradle, **as if to** get nearer **to** these objects. She tosses her hands about in the air as **if** trying to catch objects that are flying away.'"[1]

Here are some more facts which, although indeed belonging to a rather more advanced period than that treated of in this **book, are** none the less interesting to students of infant **psychology.**

**The most habitual form of** intellectual aberration **in children** appears to be mania, **that** is to say, **incoherence** and delirium **of** ideas, furious **agitation or** tranquil wandering **of** the thoughts. On this point most authorities are agreed. Dr. Delasiauve,[2] Dr. Paulmier (in his thesis **entitled** "Mental Affections in Children, and Mania in particular"),[3] and Dr. Morel,[4] declare that insanity in children generally takes the form of maniacal excitation.

The following are some remarkable cases collected from the writings of mad-doctors. We **will** quote first the very circumstantial case given by **Dr.** Chatelain, who has himself had occasion to study a child of four years and **some** months, the daughter of a farmer of the Jura. Two causes especially,—the one entirely physical (viz., measles), the other moral (a great fright caused by the sight of a fire-engine),—had acted on **the feeble** constitution of this child, **and** determined the peculiar state from which she was **suffering.** "Louise," Dr. Chatelain states, "is '*drôle*,' singular, and *distraite*. She **answers at** cross-purposes to the questions put to her. **One** day her father told her to bring him her doll; she went to fetch it, but brought back nothing, though saying at the same time, 'Here it is'; the hand and the arm made the gesture of a person giving

---

[1] Quoted by M. Compayré, loc. citat.
[2] See *Annales Médico-psychologiques*, 1855, tome i., p. 527. *Forme Maniaque spéciale chez les Enfants.*
[3] *La Thèse de M. Le Paulmier*, 1856.
[4] See *Annales Médico-psychologiques*, 1870, tome ii., pp. 260–269.

something, but the hand was empty. Since her illness her character has sensibly altered; she has entirely lost the timidity natural to her age. In the presence of two physicians who were unknown to her, and who examined her, she showed no shyness and no awkwardness. If asked a question, she answered vivaciously and without hesitation, but quite at random." And then follows a full-length description of a conversation testifying to the complete disorder of ideas in a child otherwise very intelligent, and whose ailment cannot be confounded with idiotcy.

The preceding example gives us a case of calm and tranquil mania. The little girl in question, however, was also subject to fits of furious mania, which took the form of a need of perpetual movement, of tears and screams, and threats of death to her parents. At other times, agitation is the permanent characteristic.

With rather older children cases of maniacal madness are still more frequent. Dr. Morel cites the case of a child five years old, who, in consequence of a shock caused by fear, fell into a state of "continual turbulence and maniacal exacerbation."[1] Dr. Guislain records, under the name of *furious monopathy*, a malady of the same kind in a little girl of seven years. Here the cause of the evil was a blow on the head.[2] And as moral causes always alternate with physical ones in the generation of insanity, we find in Foville a case of a boy ten years old who lost his reason through reading too much.

"It is a remarkable fact, that with all these numerous examples of infant mania, the list of which we might prolong, not a single case of monomania can be recorded. Fixity of ideas is quite as impossible in children in a state of mania, as in their normal and rational state. The little mad girl described by Dr. Chatelain was continually undergoing a change of ideas. "Generally speaking, some one idea would occupy her exclusively for a day or two, and would then give way to another." Monomania appears at first sight to be a

---

[1] Morel, op. cit., p. 102.
[2] *Dictionnaire de Médecine*, 1829.

sign of great intellectual weakness, since all the ideas and
sentiments are as it were annihilated before one single
thought, which has become sole master of the consciousness.
And yet, if we think about it, monomania presupposes a
certain force of intelligence and power of concentration,
since it is a thoroughly systematic delirium. Children, with
their mobile, incoherent ideas, their fluctuating, volatile impressions, may easily become delirious—*i.e.*, pass from one
idea to another without connection or reason ; but we can
understand their not having sufficient strength to group all ·
their faculties round one single mad conception in a permanent manner. This then, no doubt, is the reason why intellectual disorder manifests itself in children by a rapid and
incoherent succession of ideas, incessantly and distractedly
chasing one another, rather than by the obstinate concentration of all the forces of the mind in one direction." [1]

One of the most curious forms of imagination is automatism. It is one of the most important phenomena of
somnambulism, and one which will help us to realize what
goes on in a child's imagination. "If we ask a sleeping
person to say what he is thinking about, he will always
answer, that he is not thinking of anything, and that he has
no ideas whatever. We must take this answer *au pied de la
lettre*. A somnambulist does not think at all. His intelligence is a blank—absolute vacuity. This psychic inertia
shows itself by the complete inertia of the face and of the
voluntary movements. But if, in the middle of this utter
blankness, an image or an idea is suddenly presented, this
idea will at once take possession of the entire imagination." [2]

Thus in children, their relative psychic inertia explains
the vivacity of their impressions, their strong objectivity,
their temporary possession by some idea : they are, as it
were, at the mercy of momentary impressions, and especially
of those of the greatest intensity. All that they see they
believe to be exactly as they see it. All that is told to them,

---

[1] These remarks are in accordance with those which I have made in
the chapter on Memory.
[2] Ch. Richt, *Revue Philosoph.*, Oct. 1880, p. 361.

M

rises in vivid images before them. All the ideas that are suggested to them appear to them in a visible form, and they instantly proceed to execute or imitate them. All this explains the powerful influence of example on these pliable little spirits, and also the contagious effect on them of sentiments expressed by words and gestures. As the person magnetized is subject to the magnetizer, so, in a lesser degree and without complete continuity, children are subject, physically, intellectually, and morally, to the suggestions and influence of the people about them. They are veritable automatons, obeying consciously or unconsciously, with more or less entire docility, the tyrannical influence of any image that suggests itself or is presented to their minds. The following extract describes a hypnotized subject, whom the automatism of which we have spoken reduces in some respects to the level of a child.

"I say to V., 'Stroke that dog.' Immediately she goes up to the dog and strokes him. If the dog tries to escape from her caresses, V. runs after him; if he goes out of the room, she attempts to follow him. If you put an arm-chair or a bench in front of her, to stop her from passing, she either knocks it over, or, if she cannot do this, she gets irritated with it, and pushes it angrily away. But at a sign from me, she will stand still, trembling with rage and indignation. I give her some object or other, a pencil for instance, forbidding her to give it up to any one. If one of the assistants tries to take it away, she will make a desperate resistance, running across the room, fighting, biting, and kicking, in a state of indignation which one could hardly form an idea of if one had not witnessed such scenes. Professional magnetizers take pleasure in exhibiting such scenes in public, and sceptics decide that they are impostors. This they might certainly be, for there is not one of the phenomena which cannot be counterfeited. These performances, however, are not necessarily trickery, and nothing is more real than the complete subordination of all the intellectual forces to a word of command. It seems as if the sleeping or mesmerized subject had no other concern than to conform to the intimations received. One idea has taken possession of the intelligence and has become sovereign master. All beside is as nothing. All is darkness by the side of this one vivid idea, and everything which hinders its execution is repulsed with anger."[1]

---

[1] Ch. Richet, *Revue Philosophique*, Oct. 1880, p. 365.

# CHAPTER X.

## ON THE ELABORATION OF IDEAS.

### I.

### JUDGMENT.

" IF intelligence cannot exercise itself without distinguishing and comparing, neither can it do so without affirming, either explicitly or implicitly, verbally or otherwise. In other words, every intellectual operation presupposes a judgment ; for judgment, according to the definition of Aristotle, is an operation which consists in affirming something about something. We cannot declare anything without believing it; and to believe, to affirm, to judge, are all one. What is it to perceive an object? It is to cognise its shape, position, distance, dimensions, etc., etc. To know, for example, the distance at which an object stands from us, is this not to bring to bear on this object, mentally at any rate, a certain affirmation? Can we form a general idea, without affirming that this idea extends to such and such individuals, and comprehends such and such qualities? We see, indeed, that imagination makes us conceive fictions or chimeras ; but even then we affirm either the possibility of the things which we conceive, or the desire that we have to see them realized. In short, we affirm the subjective fact of these dreams, even when we do not think that they correspond to anything external. Thus, in a word, we can neither perceive, nor compare, nor abstract, nor generalize, nor remember or imagine, without making an affirmation or a judgment." [1]

---

[1] Joly, *Cours de Philosophie*, p. 94.

The above is a psychologist's definition of judgment. A physiologist will perhaps explain, up to a certain point, its origin and working. Judgment, according to M. Luys, is only the reaction, the repercussion, the affirmation of our personality "in the presence of an excitation from the external world, either moral or physical. The action of judging, so far as it is a physiological process accomplished by means of the cerebral activities in movement, is decomposable into three phases, which are as follows :—

" 1. A phase of incidence, during which the external excitation impresses the *sensorium* and rouses the conscious personality to action.

" 2. An intermediate phase, during which the personality, seized upon and impressed, develops its latent capacities and reacts in a specific manner.

" 3. A final phase of reflection, during which the process, continuing its progress through the cerebral tissue, is projected outwards in phonetic or written co-ordinated manifestations. The impressed human personality, in fact, expresses itself, exhales itself in its entirety, in either articulate or written language." [1]

This interpretation accounts both for the diversity of human judgments and their greater or less rapidity in operating, and also for the necessity of these judgments which are common to the species—principles of common sense, which correspond to whatever there is in common in all cerebral organizations.

If to judge is to believe something about something, it is evident that little children, like young animals, are conscious of their sensations, that they localize them elsewhere than in some part of their own persons, that they appreciate and differentiate them, that they know themselves without having seen themselves, and know their mothers—in a word, that they judge, *i.e.*, they believe something about something.

We have only to spend a few hours with an infant of two or three months, to convince ourselves that it is capable of

---

[1] Luys, *Le Cerveau et ses Fonctions.* See Eng. Trans. (Internat. Scient. Series), p. 289.

forming judgments. The little Tiedemann who, on seeing his nurse take up her cloak, guessed that she was going to take him out, and instantly showed signs of delight, went through several acts of judgment or belief. He believed that the person who approached the chest of drawers, raised her arms and took out her mantle, was his nurse; he believed that she made all the movements just stated; he still further believed, by virtue of familiar association, that these acts would be followed by others equally well known—the going out of the house, the ride in his perambulator between long lines of houses, amidst the noise of the carriages, horses, and dogs, the sitting under the trees, etc. . . . All the things which he believes to be and which he sees, all the other incidents which he believes will happen because he has generally seen them follow the first, are modes, manners of being, forms of persons and things which he remembers, and imagines either as co-existent or as following each other, they are so many sympathetic judgments à *posteriori*, as Kant might have said.

When little Marie, at three months and a half old, distinguishes the various parts of her body, plays with her mother and strokes her face, chatters to the flowers, stretches out her eager little hands towards them, utters admiring exclamations at the sight of their brilliant colours; turns round towards the bird-cage if her mother says, "Listen to Coco!" understanding the gestures, the tone of voice, the faces of those who speak to her, laughs when she is praised and cries when she is scolded; turns away her face dejectedly so as not to see another child on her mother's lap, etc., etc.; is there any need to demonstrate that all these facts imply judgments of a very characteristic nature?

Thus, in the case both of little children and young animals, the only debatable point is the degree in which they possess this faculty of judging. They do not form abstract judgments as grown-up people do, but they go through the same mental operations as adults at the sight of anything—a person, an animal, fruit, toys, a piece of furniture, etc. They recognise these objects as having been seen before, they discern in them such and such qualities, they distinguish them **from the** surrounding objects, they discover in them either a resemblance or a difference, they range them in such or such a group ; and all these operations presuppose **a** judgment, or are in fact themselves the judgment.

If all intellectual operations, even the most simple, imply in a certain measure the operation of judging, it is the association of ideas, added to the **mental** grasp of the individual, which has most influence on **the** formation and nature of the judgments. " Two minds, constituted exactly alike **and** placed in exactly the same surroundings, would **develop in the same manner and** would produce **the same thoughts. . . . In the** generation **of an idea** there **are two sorts** of causes ; the state of the organism, as dependent upon anterior impressions, and this same state as affected by physical conditions.

Now these two kinds of states vary in different individuals. The regions of the brain where ideas are elaborated, are not equally abundant in nervous elements ; and these elements are not endowed with such strong vitality, are not **so impressionable,** and do not combine so well in one individual as in another. Hence all the different degrees of brain power and aptitudes which show themselves in differences of judgment.

Three little children, from eight to nine months, are seated at play in the middle of the room. I speak very loud to attract the attention of all three, and at the same time I **place at** a distance of a yard from where they are sitting a **little white** horse on wheels. One of them, **A, instantly** utters a **cry of** admiration ; B opens his mouth and fixes his eyes on the toy ; C looks at it with indifference. Five seconds have now passed ; A throws **up** his arms and stretches his head forward ; B also stretches out his arms, but more tranquilly, and as if in imitation of A ; C looks on **at the** agitation of the other two, and then in his turn utters a sort of warbling cry. Four more seconds have passed. A can contain himself no longer ; he precipitates himself forward, rolls over on his side, knocks his head against the floor, and begins to cry. I help him up again, whereupon he instantly recommences the same gestures and movements, and tries to crawl on all fours towards the horse ; B up to this moment had never taken his eyes off the toy, but he now looks at me, makes beseeching gestures, and is on the point of crying.   As for C, he looks at A going on all fours, and seems more engrossed with this than with the horse.   This example suffices to show how the aptitude for judging differs—both as to rapidity and quality of judgment—according to the difference of impressionability and personal reaction.   The older a child grows, the easier it is to observe these differences ; but when he arrives at the initial period of speech, the most superficial observer cannot fail to notice them.

Whatever difference the varieties of organization and the accidents of intellectual life produce in the judgments formed by various minds, there are nevertheless analogies common to the species, and also general similarities and universal judgments, which constitute what is called common sense, or reason.   There is no human intelligence deficient in the ideas, more or less distinct, which certain philosophers still call innate or primary, and which would be better designated by the name of essential.   These categories of Aristotle, subjective forms of thought, ideas of quality, being, number, time, space, etc., etc., correspond to the structure of the brain, the constitution of the human mind, and the immutable properties of nature.   Every human being, in presence of external phenomena, cannot fail to acquire them at an early age.   We are all born with intelligences capable of deducing them with more or less rapidity and energy from our first impressions.   And thus the theory of innate ideas can very well be reconciled with that of the *tabula rasa*.   There is a reason in things which corresponds to human reason, the latter being only the product or image of the former.   Reason as a faculty for the production of pure ideas, does not exist.

We have seen that children of three months go through a process of judging similar in many respects to that of grown-up people.   We must now show by what stages they arrive at developing this faculty, which begins by being a simple affirmation of existence, passes from the concrete

to what is called the abstract, becomes comparative judgment, and extends to the conception of similarities more or
less numerous.

The first form of judgment is the apprehension pure and
simple of a concrete object with one or several of its qualities.
This object may be the child's own person, or the persons
surrounding it. We cannot say by which of these two kinds
of apprehension the child begins. It is probable, that at
first he exercises both at once. The first judgments are
nothing else than the immediate application of the child's
mind to his own being and to other objects; that is, to the
sensations which he feels himself, and feels or infers of other
things. But just as the first judgments are without distinction of subjective or objective, so they are also without
distinction of subject and attribute. For children, as well
as for inferior animals, these terms remain for some time
fused together. An oyster has some feeling of itself, and
some feeling of external objects; it does not affirm its
existence, but it feels it and believes in it strongly. It also
believes firmly in the existence, *i.e.*, the reality, of external
objects. These are judgments of existence, modality, and
externality implicitly formed. In like manner with the
fœtus, its own existence and that of other foreign bodies
affirm themselves to it from the time of the first conscious
sensations; they are implicit and confused, but, at the same
time, real judgments. In the period which follows birth,
a child's judgments begin to be a little more pronounced,
without nevertheless going for a long time beyond the
sphere of individual concretions. This phase may last till the
end of the second month, but with a more or less marked
tendency to pass from the concrete to the abstract, from the
individual to the general. Noteworthy progress has been
made in the faculty of judging when a child begins to
recognise images, and still greater progress when he expects
certain facts or sensations to follow certain appearances.
This is what takes place in all the examples cited at the
beginning of this chapter. Primitive judgment, in its first
phase, is a discernment of perceptions ; the second phase is
an association of images, a suggestion accompanied by belief.

Judgment, properly so called, is already found in due proportion in children three months old. When a child of this age mentally applies an idea of quality to such and such form and colour of any object,—for example, when he distinguishes one kind of food from another, one animal from another,—there is already established in his mind a tendency to recognise likenesses and distinguish differences. This is a mode of judgment, very poor as far as comparisons are concerned, and absolutely void of generalities, but which none the less resembles the ordinary judgments of an adult. A child of ten months begins to have this judgment of existence, which I will not call abstract in the philosophical sense, but which is separate and distinct while remaining concrete. In fact, whatever may have been said on the subject, judgments regarding existence form no exception to the general rule. They also have for their origin, condition, and occasion, those collocations of images and representations which produce in our minds ideas accompanied by evidence, belief, and recognition.

The presence and the absence of a person or a thing are already very distinct facts, even for a child who cannot yet speak; for him, to exist, is to be there; not to exist, is to have disappeared. Hence he understands thoroughly what is meant by the words *a-pu* (*il n'y en a plus*), which he soon learns to repeat himself, as well as *peudu* (*perdu*). The judgment of existence is confused at first with that of presence or absence, and relates always to something concrete. Children are capable of this.

It would be a mistake to suppose, that in order to deal with the evidence which supports even tacit affirmation, judgment needs the assistance of very wide generalizations. A very distict image, which the child connects with the representation of an object, has for him the same force of reality that it would have for an adult—it is to him reality itself. It is not necessary that the conception of always, nor the whole number of known similarities, should be added to this evidence; a few well-grounded experiences, a few similarities, or even one very distinct one, will suffice to constitute judgment in its fullest import. Thus we see that

minds which are limited in their ideas, which contain few
images and few associations, are the most thorough and
often the most ardent in their affirmations.

If we look carefully into the matter, we find that generality
of thought carries with it, if not a certain vagueness, at any
rate an amount of indecision, which may have the appearance
of breadth of judgment, which is often the precursor of it,
but which is a necessary stage before it becomes an intellectual merit.    The more the power of generalizing is
extended, and the more numerous are the known examples
of similarities, the more they suggest ideas of real differences,
and consequently of possible affirmations.    Thus the *savant*,
or the man of experience, will hesitate and suspend his
judgments and actions, where ignorant and inexperienced
people, adult animals, and children, will judge and act
precipitately.

It is because we are ignorant of the natural course of
development of a young child's mind, that we imagine his
logic to be at fault when it is in opposition to that which
we practise, and the principles of which we inculcate on
him ready-made in our language.    As our judgments relate
to comparatively general conceptions, at any rate more
extended and more numerous and quite different from those
of a child, the words which he uses have not the same
arbitrary signification which we attribute to them.    To him
they represent facts, whereas we make them signify collec-
tions of similarities, general affinities, etc., etc.    He applies
one of his general ideas, that is to say a particular and
empiric term, to a given sensation, but as our interpretation
of his judgment is not a relation between two facts, but a
relation between an idea and a fact, we find the statement
false.    It is truth to him, but error to us.    Those who
observe and bring up young children ought to be convinced
that the progress of thought at the earliest age is in relation
to that of the faculty of analysis bearing on objects, rather
than to the progress of speech.    Too often, it is true, we
exaggerate this fact, and imagine, like Condillac, that infants
are capable of observing " down to little details even," not
external objects only, but even the operations of the mind.

According to this old theory, but lately still in repute, children, while learning to express their judgments in words, also learn to analyse them bit by bit ; the art of thinking is, so to say, the art of speaking.   Yes ; if children spoke their own language, and not ours.   Condillac, however, by a slight contradiction of himself, admits that children judge and reason before they have learnt to speak.   In truth, if abandoned "to the sole impulses of nature," children fall into frequent and serious errors.   Animals, which some people look upon as a raw product of nature, often make the greatest mistakes, even in things which concern their own interests.   The fallibility of instinct is a fact ascertained by science.   But in the case of children, in addition to the errors which proceed from the spontaneous use of their own faculties, there are a great number which we must attribute to the mistaken meaning which they attach to our language, and to our no less mistaken interpretation of theirs.

The first business of developing infant faculties is carried on in great measure by nature alone ; and her method is by exciting cravings and desires, which are the condition of useful·exercise of the faculties.   A child learns chiefly because he finds pleasure and use in it ; he acquires distinct ideas through the need he feels of disentangling his thoughts. He acquires them for himself, by the simple method in which he uses his senses.   But do not his mistakes arise from the urgency of his desires and needs, which lead him to form hasty judgments?   This may happen, and indeed does happen often.   He has then used his senses badly ; but the error may be only momentary.   "Deceived in his expectation, he finds that it is necessary to make a second judgment, and this time he judges better ; experience, which watches over him, corrects his mistakes.   Does he think he sees his nurse, because in the distance he sees some one who resembles her ?   His error does not last long.   If a first look has deceived him, a second undeceives him, and he again uses his eyes to seek for her.   In my opinion Condillac greatly exaggerates the judicial power of children,—especially of quite young ones,—and their power of reflection and of **reaction from their first** decisions. This correction of the errors of sense by fresh observations is performed by children unconsciously; they judge and **un**-judge quite mechanically.

What also appears to me extreme in this theory (adopted in part by Rousseau and reproduced with *éclat* by H. Spencer), is to suppose that young infants perceive clearly " that pain follows a **false** judgment, just as pleasure follows **a true one.**" It is **true that by** the time children are five or **six months old (but not at** the beginning of their existence) **they will already have** acquired, though in a very limited degree, **the** power of understanding and utilizing **some of** the lessons of experience. This is quite **true.** But that pleasure and pain " their first tutors " **give them** infallible lessons, " because they teach them **whether they** are judging rightly or wrongly," is an untenable theory. Pain and pleasure are indeed teachers **of most** incontestable authority, but themselves also requiring **to be** trained. It would be equally hazardous to hold absolutely by the conclusion which springs naturally from these premises, that during infancy we make rapid and astonishing progress because **we** make it under the direction of these marvellous teachers **of wisdom.** The inherent susceptibility and retentiveness **of the** organs in infancy, when they are so elastic and almost free from impressions, the very nature too of the impressions **at** this period, relating as they do chiefly to the most essential **needs** and being comparatively few in number and easy to **discern, in short,** the preponderance **of** the power of ab-**sorption over the power of digestion and** combination, count for something **with the** teachings of nature in this rapid but superficial progress. It must also be remembered that a child scarcely a year old, whose faculties **and** claims **have** increased, finds himself in presence of a vaster and more difficult work, and it **is thus** natural that his progress should **appear** proportionately small. Besides all these causes **of error,**—which are natural because they result from the nature of things and from that of the mind,—and those special causes of false judgments which result from particu-lar organic predispositions **or** psychic constitution, another important and constant one is the necessary imitation of our language and intellectual processes.

Nature always gives serious warnings with a view to rectifying the numerous errors committed in spite of her, or even with her aid. But these warnings are specially intended for adults, and intelligent mothers understand how to introduce them into their training. I knew one mother who, in order to prevent her child from going near the fire, allowed him to burn himself slightly with a hot iron which she held out for him to touch. The child thus made this dangerous experiment for himself, but at the same time safely under his mother's eye. He cried; and his mother comforted him and said to him several times, while showing him the iron: "It is hot, it will burn Edmund." This experiment, repeated with variations and always in the same safe manner, taught him very quickly, and by the time he was twelve months old, to be afraid of the fire and of any objects that were taken off it or out of it. From hearing these two words, *hot* and *burn*, and seeing the simple action of blowing, which is both a visual and an auditory sign, applied to a few well-defined objects, he had learnt the meaning of them. He even acquired the habit of making a gesture of repulsion and imitating with his lips the noise of blowing when he saw anything near him which smoked or seemed hot. Thus by experiment of this sort, and by the simple play of association of ideas, we may lead children to form right judgments about a great many things, and to act in conformity with these judgments. Nature and the child itself carry on three-quarters of the work of natural education, we ourselves must do the rest.

By attentive observation of children we obtain a multitude of indications, not only concerning the genesis of their faculties, but also as to the direction to be given to them. For instance, children of their own accord judge by comparison; it is for us to watch over and help their judgment without seeming to do so.

A child will go on tormenting a cat or a dog till suddenly the animal turns round and bites or scratches him; and then he says: "Bad pussy," "bad doggie." It is for us then to make him correct the judgment by which he has affirmed his conception of the relation between the idea of cat or dog, and scratch or bite. When he says, "Good baby" or "Naughty baby," he again makes a comparative judgment, bearing on the relation between such and such actions and the qualification they generally receive. Our part is to place the child frequently in positions to renew such judgments, with the use of the same formula, *à propos* of actions and events of which we cannot always actually bring about the occurrence, although we can always turn it to profit. Thus, without crushing the child's initiative, we shall be doing all in our power to increase its store of observations, of terms of comparison, and of similarities—in one word, its intellectual sphere.

But here the great danger to be feared is, that we should aim at developing children, not for themselves, but for ourselves. How often we see people endeavouring to inculcate on children the judgments and the practices of adults, without stopping to consider how much their young minds are capable of understanding or retaining. I will cite but one example, of which I was a witness at a *table d'hôte* at Luchon. It is the custom in good Spanish families always to leave something on one's plate, in order not to appear too great an eater. They accustom their children to this habit by telling them frequently: "Politeness requires you to leave something on your plate; don't forget this, it is politeness." At the table where I was sitting, a little girl, two-and-a-half years old, had reluctantly left a portion of some sweet dish. She leaned toward her mother and whispered in her ear, but loud enough for every one to hear: "Please, mamma, may I eat politeness to-day?" It is thus we are understood by children whom we do not take the trouble to seek to understand. We either give them credit for too much or too little intelligence.

And here I must point out an error common enough in moralists and educators, and doubtless also in parents, viz., the idea that children are good observers and judges of character. This requires explanation. Their faculty of discernment, very limited in its scope, is always confined in rather narrow subjective bounds. Like animals and savages, they only direct this faculty of discernment towards the most salient external qualities, and, above all, those which have some interest for them, either actual or near at hand. The attention, awakened by interest, causes a child to notice a thousand differences, full of importance for himself, though often of none whatever to other people. That which generally takes hold of him at first sight, is the expression of the gestures, the physiognomy, the tone of the voice, the simple and common sentiments, the broad signification, rather than the delicate shades even of emotions which closely touch his sympathy. A child of six months knows very well with whom he has to deal (but in respect to a few actions only)—if he is with a severe father or an indulgent one, a firm or yielding mother, an uncle who is fond of playing with him, or a sedate aunt, lively good-tempered companions, or dull and surly ones. He knows very well, even at this age, but better still a few months later, what he may venture with the one and the other. In like manner, animals distinguish their prey a long way off, and proceed towards it in a very different way than they do towards other animals. In the same way too the acuteness of sense, or rather the judicial sensibility, of savages is exclusively limited to the kind of perceptions which they have had most interest in discerning; just as a compositor can distinguish by the mere touch the letters which he wants; or a seamstress can pick out with a single glance the needle and thread of the size she wants from amongst a quantity of others; or a botanist will tell you the name of a plant which, either from its distance or diminutive size, can hardly be seen by other people. This faculty of discernment is a matter of interested habit, but it is nevertheless liable to numberless errors.[1]

---

[1] As ideas scarcely exist without judgments, or judgments without reasonings, we may say, that error is generally the result of faulty induction or deduction. We shall therefore treat more particularly of error in the chapter on Reasoning.

We may sometimes notice in quite young children judgments which are as concrete and as little general as possible, which in an adult pass for general judgments. I will give an example bearing on plurality of repetition, which is almost always confounded with a definite number of times in which an event has happened. I was walking with my nephew Charles, then four years old. We were going along a canal bordered with willows and alder-trees. Charles, who was a born naturalist, and had his eye on everything, talked on without ceasing in the most interesting fashion, and suddenly showed me a dragon-fly with blue gauzy wings which had just flown away at our approach. "Generally," Charles began to say (and I listened attentively, asking myself what this fine-sounding adverb might mean from a mouth of four years old), "generally," he said a second time, "dragon-flies sit on the leaves in the sunshine, but not always." I confess I was surprised to find so young a child performing thus a process of *applied* abstraction, or at least showing a tendency towards general abstraction apropos of concrete facts, which had been interpreted before him, and of which he had understood the interpretation.

There is not a single specimen of this sort of abstraction, resulting in practical judgments, of which one cannot discover the rudiments even in a child of twenty months. But we must not surely expect them to distinguish between the good and the evil, the qualities and faults of persons and of things ; or, if they do, it is only after the fashion of so many ill-educated and not over-wise adults, without knowing it and in a very slight measure. It is however very certain that the ideas a child has of its father and mother, and all the things it knows are not simple ideas ; but that he has formed general and dominant conceptions of them—mental *wholes* from which he is continually cutting off abstractions proportioned to his judicial capacity. The people he knows but little are wholly good or bad, without modification, like the heroes of pseudo-classic tragedy. Those whom he knows well, are sometimes one sometimes the other—chiefly according to his own conduct.

N

Thus, within very narrow limits, his judgment proceeds from the abstract to the general.

## II.

### ABSTRACTION.

As all our ideas are produced by sensations, we may say that each distinct idea is the result of analysis. " No one sense presents to us all the qualities which we perceive in a body. Sight presents colours; hearing, sounds, etc. . . . When we use our senses separately, the objects become decomposed as it were : we observe successively the different parts of a watch. Touch is, of all the senses, the one which reveals to us the greatest number of qualities ; but though it presents several of them synchronically, we only note them successively." [1] This necessity of analysis, inherent in our organization, may be considered as having two principal degrees or forms, that which results in notions of individuals, and that which results in notions of modes or qualities, or abstract ideas, as it has been agreed to call them. We shall study both these forms of analysis in the young child.

The most important stages in a child's progress take place when the senses of sight and hearing have come to the aid of touch and taste. It is at this period that we can begin to take hold of some of the manifestations of the young intelligence in process of development. We should form a false conception of the first visual perceptions of a little child, if we compared him to the blind youth whom Cheselden operated on for cataract, and who saw different objects placed before his eyes only as masses of colour spread over a plain surface. In the first place the sense of touch has not yet furnished the child with the conception of a plane surface, and secondly, the field of vision only opens out to him gradually. Tiedemann says of his little son, when only two or three days old : " His eyes already moved in all directions, not by chance, but as if he were looking out for objects, and they seemed to rest

---

[1] Condillac, *Cours d'Etudes*, t. i., p. 47. *Edit. Dufart, an II. de la Rép.*

by preference on things in motion." This observation is
not strictly accurate. I have never seen any infant before
the age of eight days follow any object with its eyes except
a candle or the flame of a fire. But I have seen many at
a month old whose eyes could follow (their head all the
while remaining immovable) the movements of an object
waved about at a distance of a few inches, or who would
fix their eyes for a length of time on any brilliant object
two or three yards off. In proportion as a child's visual
organs are exercised and its field of vision enlarges, visual
perceptions come to it as luminous or coloured particles.

In what would be a picture to an adult, a child sees only
the salient points. His eye only distinguishes the different
colours little by little, and the want of adjusting power in
his sight prevents his seizing any but a few prominent
objects or even some parts of these objects. This must
be the case if the observations which I have recorded in
the chapter on attention are accurate. First of all, it is
his mother's or his nurse's breast, the bright flame or
reflection of a candle, the outline of a piece of furniture or
a window, anything striking in form or colour (these are
everything to children) which strike his retina and engrave
themselves on his memory. External impressions seem to
make their way to a child's eyes, as they do to all his other
senses, bit by bit, colour by colour. Hence, to perceive
sensations distinctly, to preserve a distinct recollection of
them apart from the vague complexity of the concomitant
impressions which have only very slightly affected the senses,
this is a work of separation which may be considered a
sort of rudimentary abstraction.

With these primary perceptions, parcelled out and isolated,
which we look upon as the first abstractions, there mingle
by degrees secondary perceptions, the whole sum of which
make up the first concretions. Now one of the dominant
perceptions will recall to the child and lead him to infer
a whole group of secondary perceptions associated with
this dominant one. Such and such a colour or sound, or
any particular tactile or sapid impression, will instantly
suggest the idea of such and such a form, object, or person.

The person of his nurse will at first have become known to him by a continual succession of analytical perceptions. Her breast, which afforded him the keenest enjoyment, and against which he nestled his little face, delighting in its softness and warmth ; her warm caressing hands, so sympathetic both to touch and sight ; the different parts of her face, her eyes, her cheeks, her lips, whence come the coaxing words and pretty songs, and the smile he so soon learns to understand. The different parts of his cradle, the bars and legs of the chairs, the polished surface and shining edges of the table and the surrounding pieces of furniture, are all objects from which there have come to him successions of detached impressions which will presently combine among themselves to form individual conceptions. But these ideas of individuals and whole objects are still extremely vague and incomplete. I knew a child of two months old who could clearly distinguish a person from an animal, or from a piece of furniture ; but he used to smile indiscriminately at the first comer, and would seek the breast of any woman who took him in her arms. But at three months he could so well distinguish his nurse from his mother that if, when his mother was holding him, his nurse took another child on her lap, or he saw her being embraced by any one, he would at once show his jealousy by frowns and tears. At this age he also clearly distinguished a cat from a dog, the former having scratched him more than once, whereas the latter overwhelmed him daily with caresses ; the moment the dog appeared, he always showed great delight. One month had thus sufficed to fix clearly in his mind a large number of individual conceptions.

The necessity which children are under of seeing in a detached and scrappy manner in order to see well, makes them continually practise that kind of abstraction by which we separate qualities from objects. From those objects which the child has already distinguished as individual, there come to him at different moments particularly vivid impressions. Thus, I place a beautiful rose, in the light of a shaded lamp, before a four-months-old child : his mother holds him upright on her knees before the table

where I have placed the pretty flower; the child opens
his mouth, babbles, holds out his arms, jumps, throws
himself forward; his mother then places him on the table,
and he twists and twirls himself about in the most curious
fashion, looking all the time at the rose.   Is it not probable
that this colour, which appears to his eye without any precise form, and which has so keenly excited his attention,
will remain in his memory as a very vivid colour, without
any definite concrete embodiment?   A few minutes after,
I lifted up the shade, and the strong flood of light, very
brilliant and soft, made him utter cries of joy and jump
about more vigorously than ever; the rose is forgotten—
he does not even see it—he is wholly taken up with
the lovely golden flame, which fascinates and dazzles his
sight.   We may be sure that the memory of this bright
and flickering flame will be preserved with far greater
vividness than that of the nearer surrounding objects,
which have only produced very feeble and indefinite
impressions on him.   Dominant sensations of this kind, by
their energy or frequency, tend to efface the idea of the
objects from which they proceed, to separate or *abstract*
themselves.   Sometimes even in their waking hours, but
more especially during their dreams, when there are no
external impressions to renew the chain of usual associations,
there is no doubt that children's brains are traversed by
impressions as vivid in themselves as they are vague and
undetermined, and that hallucinations corresponding to a
sound, a colour, a tactile, muscular, or thermometric
sensation, a form, a dimension, or a taste, awaken vivid
sensations in them without being accompanied by those
feebler sensorial impressions which were grouped around
them; *i.e.*, without the aid of any definite object.   Here we
see reproductive imagination assisting in the operation of
modal abstraction.

These acts of abstraction, resulting in individual or modal
conceptions, which the infant practises very early, have a
continual tendency to relapse into the confused masses,
the indistinct wholes, and the various concretions from
whence they were drawn.   The series, I will not say of

impressions, but of habitual conceptions, regular and natural, is that of synthetic vision for the child as well as for the adult. The individual forms are so many and so varied, that the memory allows them little by little to fall back into the great objective mass, vaguely but directly seized, and from which they only detached themselves under exceptional or temporary circumstances. The qualities separately perceived in these entities, however striking they may have been at first, are not always so in the same degree, cannot always equally captivate the attention or awaken the memory. The flame of a candle is not always equally bright or flickering; tactile, sapid, olfactory, and auditive impressions do not always strike the child's sensorium with the same intensity, nor during the same length of time. This is why the recollections of individual forms, although strongly graven on their intelligence, lose by degrees their first precision, so that the idea of a tree, for instance, furnished by direct and perfectly distinct memories, comes back to the mind in a vague and indistinct form, which might be taken for a general idea. In the same way, the essential, constant, and necessary qualities which analysis has discovered in different objects, are preserved in the mind with indelible marks; but though the remembrance of these qualities may very often be unaccompanied by any series of distinct impressions, there is always mixed with it a vague and confused image of certain perceptions which were associated with them.

Thus, however much we may try—and it is the same, I take it, with more than one psychologist—it is impossible for us to think of any abstract idea of white, red, a sharp or flat sound, beauty, ugliness, goodness, vice, number, space, extent, without imagining some particular object, more or less determined in form, to which the abstraction which we have conceived mentally joins itself, and thus becomes a concretion.

This then being our conception of the process of abstraction which Locke, Condillac, and so many others only regard as a consequence of language, it will not seem surprising that we should allow this power to exist in young children. Language up to a certain point fixes and defines, but it does not engender what are called abstract ideas. The most abstract idea that we can conceive is equivalent to the conception of the most abstract sign *i.e.*, to a *material* idea—to a kind of simplified and refined concrete. This means first, that such a thing as a pure idea does not exist, that it is a conception of metaphysicians and algebraists—a logical convention ; and, secondly, that relatively abstract ideas (the only ones which we admit) are even in their origin independent of the faculty of language. M. Vulpian goes even further than we do : he is of opinion that animals as well as children possess this power of abstraction.

"There are," says M. Vulpian, "abstractions relative to material things, or rather to the sensations which they produce on us. There are, for instance, the abstractions by which we form the ideas of trees, dogs, red, green, such and such sounds, etc. In a word, there are tangible abstractions, formed by the help of tangible properties. Well then ! It seems to me difficult to refuse the capacity for these abstract ideas, in a measure at least, to the superior animals, for it is evident that their memory, reflection, and reason are sometimes exercised on ideas of this sort.

" As to general abstract ideas, it seems to me quite doubtful whether animals have them—whether they have the slightest abstract idea of infinity, of time, of space, of dimension, number, connection, etc. What confirms me in this conviction that they have none of these abstract ideas, is, that I am not sure that man has them himself. We are here at the mercy of an illusion which has been too little insisted on. Animals, or at least, certain kinds of animals, have a sort of language which enables them to communicate with each other, by modulations and variations of the voice, or by other special sounds, or by graduated contact. Concerning this latter mode, we know nothing more remarkable than the antennal language of ants, described by Huber. But, definitively speaking, these modes of language are very different to those of man. Man alone possesses real articulate language, he alone can make very varied abstractions by the aid of this language ;

he alone, I need hardly say, can also make metaphysical abstractions. But from the fact that, by means of certain words agreed upon, we have been able to designate these abstractions, we must not deduce as a necessary conclusion that man is capable of conceiving abstract metaphysical ideas. There is no possibility of the existence of any ideas unless there is the possibility of more or less distinct intellectual representation of them. For instance, can we represent to ourselves in an abstract manner time, space, etc.? If I am not mistaken, so-called general ideas do not exist in reality, and they ought only to be considered as algebraical abstractions, so to speak."

Thus, between the abstractions produced by the adult and those which the child and the animal form, without the aid of vocal signs, the difference is not qualitative, but only quantitative.

If we take a child at the moment when he is beginning to use language with tolerable ease, we see that, in spite of the signs he gives of deliberation and efforts to remember, he is always, as he will still be when grown up, indolent as to abstraction which does not rest on an objective representation, sufficiently determinate. A child of two years perfectly well understood the sense of these phrases : "This glass is larger than this stopper;" "Baby is a good boy;" "The dog is naughty." But at three years of age he did not understand these expressions : "The size of that house," "The goodness of papa," "The naughtiness of the dog;" notwithstanding the resemblance of sound between the abstract words and the corresponding adjectives. A little girl of twenty-three months could recite fluently the names of the principal colours, but she could only identify a few of them on objects. Her father pulled out one book after another from his shelves, and showed them to her, saying: "What colour is that?" "It is white," "It is black," "It is blue," "It is red," she replied, naming the colours quite rightly. But there was one colour which she could not determine. "It is not red nor blue," she said, after several seconds' hesitation. Her father insisted : "What colour is it then?" She answered : "Not red, nor blue, nor white, at all, at all." She could not find the word yellow, although it was well imprinted in her memory, and the colour yellow was also well known to her and clearly differentiated in her mind; for she added presently : "Like little girl's hat." Her hesitation proceeded more from the difficulty of evolving an idea relatively abstract, than of recalling a well-known word. In fact, another day, she made a mistake about green, not being able to name the colour of a green ribbon ; and another time again about yellow. Her father having said to her, "Go and look on my table for a yellow paper," she brought three of different colours, and none of them the right one.[1]

We have already shown, in the chapter on association, that young infants possess the rudiments of the ideas of cause, place, time, etc. We may also observe in them, among other signs of judgments resulting in what we call common-sense notions, the existence of the notion of quantity, and in particular of numeric quantity. A very young baby, if we offer it a large piece of cake or fruit, together with a smaller piece, will stretch out its hand to the larger piece. In like manner, a kitten two months old to whom I had thrown a piece of meat, left this piece to go after a larger piece which I had thrown to another cat. As to ideas of number, children, and no doubt animals also, confuse them for a long time with ideas of greater or less quantity. I have been able to prove that a three-months-old kitten had only the vaguest notion of number. The last time but one that she presented me with a family of kittens, I only kept one of them; in her grief at losing the others, she used to leave this one alone for hours, or else carry it about from place to place ; it died when ten days old, just as its eyes were opening, and there was no doubt that its death was caused by want of care and food. The next time, I kept two out of five. She was perfectly happy with these two, and did not trouble herself about the rest; and I could not but conclude that *two* represented *many* to her, just as well as *five* did.

---

[1] Vulpian. *Physiologie du Système Nerveux*, pp. 911, 912.

M. Houzeau does not deny this faculty to certain animals, but he reduces it to very narrow limits. "There is no doubt," he says, "that certain animals are able to count, whether it be a number of objects or the number of times that a similar action is repeated, provided that the number is not high. This is the case, for instance, with the magpie (*Pica caudata*). When this bird is watched by a party of sportsmen, it does not stir till they begin to move away. If they go away one after the other, the bird will wait till the fourth has gone; but beyond this number it seems to lose count, and sometimes flies out of its hiding-place too soon, showing that it has made a mistake.[1] The following anecdote proves that mules can count up to five at any rate. In some of the cities of the United States many of the trams are drawn by mules. This is especially the case at New Orleans, where mules are preferred to horses. The St. Charles Street line has a short branch-line, that of the Napoleon Avenue, where each mule makes the journey five times before being unharnessed. The veterinary surgeon of the line, the clever Dr. Louis, one day called my attention to this fact, which I was able to verify, that the mules on duty remain silent during the first journeys; but at the end of the fifth, as soon as they arrive at the station, they begin to neigh, knowing that it is time for them to be unharnessed.

"There are some birds and quadrupeds that are capable of appreciating number, at least up to four or five. How do these animals, who have no conventional language, succeed in counting up to five, and perhaps higher? They must have a certain numerical medium, a means of distinguishing among their recollections. Is it by visual memory and the juxtaposition of similar images, as we might count by placing in thought a number of trees in file, or again by putting counters in a line? We cannot decide this question."[1]

According to the same naturalist, "Children at first can

---

[1] Houzeau, *Facultés Mentales des Animaux*, t. ii., p. 207.

only distinguish **between a** single object and plurality. At
eighteen months they can tell the difference between one,
two, and several. At about three, or a little sooner, they
have learnt to understand one, two, **and four—two** times
two. It is scarcely ever till a later age that they count the
regular series one, two, three, four ; and here they stop for
a long time. In the Brahmin schools, **children in** the first
class are only taught to count up to four ; **in** the second
**class they go up to twenty.** In Europe it **has** been found
**that** it takes six or seven years to get a child up to a hundred. **Children can** no doubt repeat the numeration **tables**
**learnt by heart before this age** ; **but this does not mean**
understanding numbers, **or** being **able to** determine **the**
numbers **of any** given collection of objects. The preceding
examples **apply** to European children of **medium** intelligence who are receiving the best instruction.

My personal observations with regard to little children
who cannot speak **have** not furnished **me with any indica-**
tions contrary to the assertions **of** M. Houzeau. When a
child nearly three months old to **whom two feeding-bottles**
are presented, **seizes both at once; when he takes two**
objects rather **than one, three rather** than two, when, **at**
eight months, **seeing near him two** cats resembling each
other, he **takes one for the other, and** thinks there is only one
cat, he is evidently confounding the idea of plurality with
that of quantity. A little intelligent child, two and a half
**years** old, **knew how** to count up **to twelve ; but he had**
**not a clear idea of the length of** time represented by **three**
**days.** "**I shall come back in three days,**" I said to him
once ; **and he answered** quickly : "**What** does that mean,
*in three days ?*" I then said : "I shall come back, not
*to-morrow morning,* **but** *to-morrow, to-morrow,* and again
*to-morrow,*" and he appeared **satisfied** with this explanation.
When **three** years **and three** months old, the same child
told the gardener that he was going away *to-morrow,* and
that he should not come back for a long time. " I shall
come back," **he** said, " in *many, many, many to-morrows—*
*in a year.*" This last formula had been learnt, but not
understood; the first **was** within his comprehension, and his

own invention. The same child, and I have remarked the same fact in many other children, had still greater trouble in going backwards in time, and it was only with great difficulty that, at three years of age, he learnt to understand the idea of *yesterday* and the *day before yesterday*.

Another child, at three years old, had only the most hazy rudimentary conception of the abstract notions of truth, of moral goodness, and even of physical beauty or ugliness. " What is truth ? " he asked his aunt, who had said to him : " Come, Charlie, tell me the truth." " It is when one does not tell lies," she said. He knew very well what telling lies meant, having been once severely punished for saying something for fun which was not true. " Being good," meant for him to be petted, not to do things which caused him to be scolded, which made his mother look sad and his father speak angrily and slap his hand. To be naughty, on the contrary, was to do all these things. But he confounded the idea of ugly and naughty. When two and a half years old, he used to repeat over and over again : " Such a person, such an animal, is very ugly ; " sometimes, but very rarely, he would say, Such and such a person or thing was very pretty. I said to him one day, in order to coax him to be quiet whilst he was being dressed : " You are a nice little boy, you are very good and very pretty." He answered, " Oh, no ! I am not pretty ; mamma told me I was not pretty." " Then are you ugly ? " I said to him. " Yes, sometimes, when I give papa or mamma trouble." He knew the difference between pretty and ugly, but he confused pretty with gentle and good, and ugly with naughty or disobedient. He had not yet got a clear idea of moral goodness or badness, moral beauty or ugliness, but he had very distinct ideas of physical beauty or ugliness. He would look at some people with more attention and pleasure than at others, because they had prettier or pleasanter faces. But we have now said enough concerning the existence in little children, in a rudimentary and concrete state, of those ideas called rational, which are both the result and the basis of the most ordinary and most generalized judgments of adults.

In conclusion, if it is not quite accurate to consider

abstraction as a special case of attention operating on concretions from the lowest to the highest rung of the ladder, let us at least admit that there is an easy kind of abstraction and a more difficult one; one kind more obtuse and more frequent, another subtler and rarer; in short, that there are degrees in abstraction. An individual idea is a synthesis in respect to the totality of its parts or qualities; it is analytic as regards the sum of individuals. The ideas of red, blue, or white are abstract in respect to the ideas of the flowers possessing these qualities of colour; the idea of colour or of number generally, and independent of any coloured object or numerical collection, is still more abstract, etc. In children we only have examples of the lowest form of abstraction. The starting-point of this faculty is no other than the faculty of discrimination. A child who distinguishes a certain number of persons and objects—his mother from his nurse, the dog from the cat, the table from the book-case, has already made abstractions. Every time that, for one reason or another, he seizes in any object one point of view, one concrete quality which stands out in his eyes from all the others, he makes an abstraction. Abstraction of this sort affords a child intense and absorbing sensations. If a candle is brought near him, his attention is fixed for a moment on the flame, which is the most brilliant part; if several objects of different colours are shown him, he looks at the one whose colour strikes his eye most vividly, so that he gets an idea of the colour rather than the form of the object. This is the sort of abstract ideas which are most familiar to young children. They are detachments from concretions. The proof of this is the difficulty that children,—and often, too, adults,—have in retaining ideas of colour, sound, form, and still more of quantity, goodness, wickedness, beauty, ugliness, etc., when the objects whence they have abstracted these ideas are no longer in their presence. Even language, notwithstanding its power of recalling the ideas of objects and qualities, is not equivalent to the presence or the distinct recollection of the object itself for reminding the child of the particular quality belonging to it which impressed him most.

We may say that abstraction is a tendency to separate which is always ideal and instantaneous; and it is by associating these abstract ideas,—which for a moment presented themselves as it were alone to the mind,—with some concrete idea, that memory will recall the ideas to the child. Children understand very well, after their own fashion, the words : " Papa good," " Baby nice," " Baby naughty "; but they attach no meaning, during the first two years at least, to the words goodness, wickedness, ugliness, and many other familiar ideas, which moreover very few adults would explain without the intervention of concrete ideas.

It is not therefore accurate to say that children shrink from abstractions; on the contrary, they understand and like them when they cost them neither labour nor mental effort—*i.e.*, when they are the least abstract of abstractions.

" Children would be quite content only to look at things as they pass. . . . We must not, however, exaggerate their indolence in this respect ; it is not so great as people think. All effort, in fact, has its charm ; abstraction, moreover, quickly makes up to us for the trouble it costs ; it produces those simple ideas which the mind above all delights in, that clearness which is its chief good. . . . Abstraction, then, is not an enemy ; quite the contrary ; and the mind becomes disciplined to it. This is why it is well not to spare children the trouble of thinking under abstract forms, so long as we do not fatigue them, and disgust them with study."

These reflections, which apply to children generally, must necessarily, in some measure, apply to children during the period in which we are studying them.

## III.

### COMPARISON.

To clearly distinguish two individuals, alike or different, as being two, is not to compare, but it is a step towards comparison. To be effectual, comparison,—or the fixing of the attention successively on two or several objects, or on two

or several portions of objects,—must be combined with
modal abstraction. It must lead to the conception of a
relation, either of resemblance or of difference, cognised
between these various objects or parts of objects. Comparison, properly so called, is impossible to very young
babies, at least for several weeks; and it only becomes
possible when their intelligence is sufficiently developed to
combine a tolerably distinct vision of details with a power
of attention sufficient to take in several objects one after
another in a short space of time. A young baby has not a
precise idea of the relations between things, and does not
look out for them. If comparisons are formed in its mind,
it is of themselves that they arise; the terms come together
in his mind and remain associated, under the form of integration and disintegration, of concrete resemblance and
difference; but the child does not complete the work
begun by chance, he does not perceive the bearings precisely, and he draws no conclusions.

A little child at two months old hesitates to take sugared
water in place of milk; he will refuse plain water, and will
get angry, scream, and gesticulate wildly after having
swallowed a draught of bitter medicine. These are all so
many different sensations for him; but how many analogous
experiments must be repeated on a large number of objects,
before he gets a clear idea of the quality which makes the
difference, or begins to examine if such and such an object
has or has not this quality. In like manner with young
animals, if several bits of meat are thrown near a puppy or
a kitten two months old, they will take any piece indifferently—probably that which is nearest; but when a little
older they will instinctively choose the largest piece, because the strongest visual impression determines the
strongest desire. But when they are a good deal older, if
these experiments have often been repeated, they will have
acquired distinct ideas of size and even of superiority of
taste; and it will now be deliberate reflection that makes
them run after the largest piece and fight for it with their
companions. They become progressively apt at comparison.

At three months old children seem to apprehend a large number of resemblances and differences; but as yet to compare very little, if at all. As I have related above, I once offered a little girl of three months first a feeding-bottle filled with milk, and then an empty one; she lifted the second to her mouth with as much eagerness as she did the first. I then offered them to her both at once, and holding them close together, so that she could see the white liquid moving in the one. She did what Buridan's celebrated ass would perhaps have done, had he had two hands, and been placed between a bucket of water and a peck of hay; he would have seized them both. The little girl, as well as her awkward little hands would let her, seized both the bottles at one grasp. Soon, without letting go the full bottle which she held in her right hand, she lifted the empty one with her other hand to her lips. Was this the result of mere chance, or because the empty bottle was the easier to lift? It matters little to the question with which we are concerned here. She made several fruitless attempts at sucking, and then pursed up her mouth and began to cry, letting fall the unsatisfactory bottle. Soon, however, she became aware of what she had in her right hand, applied it promptly to her lips, and sucked in several drops of its contents with evident satisfaction. Up to that moment the child had seen no difference between these two objects; and possibly even, as her attention had only been superficially directed towards them, she had seen but one object in these two so nearly alike.

I wanted to vary my experiments on her; and knowing her to be very fond of brilliant colours, I placed before her in succession some engravings of different kinds. I began with pale tints. The child, on seeing them, started forward and uttered some joyous exclamations, and stretched out her hands. But her joy was beyond bounds when I placed more brightly coloured ones before her; she felt them all over, thumped them with her hands, rubbed and crumpled them up, put them in her mouth, and gazed at them in ecstasy. I varied the experiment a second time. After some minutes devoted to another amusement, I placed

two pictures before her, one of which was brilliantly coloured.
She threw herself towards both, as if they were one single
object.   I then showed them to her separately ; she evinced
great pleasure in looking at the more sombre one, but with
regard to the other her joy and excitement were indescrib-
able.   But notwithstanding that she had been so strongly
impressed by these striking differences, they had been
nothing more to her than more or less vivid impressions ;
there was no reversion of the attention to the objects which
had produced the impressions, not the slightest suspicion
of difference of quality, or even of quality possessed in
diffۥrent degrees, not the slightest attempt at comparison.

From similar experiments I have come to the conclusion
that comparison, as we understand and practise it, is not
possible to a young brain of only three months.   Is the
power acquired a little later ? and, if so, in what measure ?
Here again I leave facts to speak for themselves, without
hastening to draw conclusions.

Some years ago, I had two grey cats, mother and son,
alike in shape, colour, and general appearance.   One day,
a child of eight months old who was brought into my room,
perceived the she cat at a little distance from him, and
called out for it with expressive gestures and sounds.   I
called the animal up to me, and friendly relations were soon
established between it and the child.   The cat raised her
head and her back to be stroked, purred, and rubbed itself
backwards and forwards against the child's frock.   The
child squeezed the cat in both his hands and pulled it
about to his heart's content.   While this was going on, the
other cat came on the scene, approached the child, purring,
and demanded his share of the caresses.   What was the
child's amazement at seeing on his left hand the same
animal that he saw on his right.   He turns on one side
and sees a grey cat, he turns on the other and sees a grey
cat there also.   He seemed as if he could not believe his
eyes, and could not make up his mind to see more than
one object, his mind being absorbed with the vivid conception of one alone.   What could have been passing in
that little brain ?   The eyes, the movements of the arms,

the bending expectant attitudes, all testified to very strong amazement. But the uncertainty of his judgment did not last long; and, before the end of the visit, having seen the two cats before him, side by side, and at a distance, and in different positions, he carried away the remembrance of two similar animals, without however saying to himself that they were alike. Experiments of this kind often repeated, —and they are repeated every day,—provide abundant materials for abstraction, and enable children to detach from similar or different perceptions the conception of analogous qualities, and stimulate them to make comparisons.

I knew a little child of ten months who made a great distinction between cakes and potatoes, which he was very fond of, and bread, which he liked less, and also between meat and bread, milk and wine, and wine and water. I subjected him to the following experiment. On one side we placed a cake, on the other a piece of bread; without hesitation he seized the cake and carried it to his mouth. It was then taken away from him, and it was a sight to see his wild agitation, the energetic movements of his arms, his face purple with rage, and his eyes all the time following with fixed attention the movements of the hand which had ravished the cake. Soon the bread was offered him; he fell into the snare and took it; but before biting it he recognised his error and threw it away with signs of violent anger. If one wants him to eat soup, one must not allow him to see potatoes, meat, or fruit. He also behaves differently with regard to his own playthings and those of other children. He has no objection to taking possession of their toys and piling them up with his own; but if any one touches his, he snatches them violently away, thus showing that he is able to distinguish his own things from others'. Also, if two or more objects are held just in front of his eyes, he sees clearly that one is not the other, and he also sees in what they differ; he compares almost without knowing or wishing it—this is a great step in advance.

This slow progress in the capacity for comparison, properly so called, has to do with the nature of the attentive power, which is so slight and fluctuating in little children that they only observe objects superficially, even those which interest them and appeal vividly to their sensibility ; and generally they only observe them from the point of view of the particular emotions they excite at a given moment. A child of three months tries to take hold of, to touch, to raise to his mouth, or to throw down anything that comes to his hand ; and, whether from curiosity or from the need of using his restless activity, one thing is quickly abandoned for another, to be again perhaps taken up in a few moments. A child of this age is relatively so ignorant that he finds very little to see in any one object, especially in inanimate objects, which have not the power of living objects to modify themselves in all sorts of ways and to keep continually alive the curiosity which they first inspired. This is why very young children who cannot yet speak are content with a cursory notice of the most prominent analogies and differences. The past being of less interest to them than the present, they feel but little need of comparing.

But from the age of fifteen months, and especially between twenty months and two years, when words are beginning to come to their aid, they begin to make large use of their faculty of comparison. They are very little on the look-out for differences, although they are very much struck by them when they see them ; but everywhere they are on the look-out for resemblances. I have even noticed this in a little boy of thirteen months. As one of his cousins was like his uncle, having the same sort of beard and the same kind of figure and voice, the child treated him at once as an old acquaintance. He called him *Toto*, insisted on being taken to him, poked his finger in his eyes, nose, and ears, wanted to sit by him at meals, found his way to him crawling on his hands and knees, as soon as he was awake, in order to play on his bed. After meals he insisted on being taken on his cousin's lap, and thence proceeded to climb on to the table, where he began all the little games which he was accustomed to play with his uncle. Seeing a pencil in his cousin's hand, he took it from him, put it in his mouth, and made with his lips the movements and sounds of a man who is smoking and puffing the smoke into the air. His uncle used to smoke. When he got down from the table, he took his cousin's hand, and said in a tone of entreaty : "*Lou, lou, lou, lou;*" this was explained to the cousin as signifying that he was to imitate the dog, as his uncle was in the habit of doing, to the child's great delight. Out in the garden, the child made another request which the cousin did not understand, much to the astonishment of the former, who was accustomed to being instantly obeyed by his uncle. Thus this child, only thirteen months old, reasoned and acted on the ground of analogy; from certain resemblances in objects he inferred others, or even complete similarity. At first it would surprise him, not to find the other points of resemblance combined with the first, and he would seek for them with attention heightened by vexation. His cousin, having been coached up in his part, humoured as far as possible all the habits which the uncle had made necessary to the child ; but some he replaced by ways of his own ; and the end of it was, that after being with his cousin for three weeks, the child afterwards expected from his uncle all the gestures, tones of voice, games, indulgences, and acts of obedience which the new *Toto* had accustomed him to. The parts were changed, but the intelligence, the sensibility, and the will of the child worked in the same fashion.

At the age of two-and-a-half children compare a great deal. They frequently use such phrases as these : "baby tree" (little tree), "papa tree" (great tree), "mamma duck," "baby duck," to signify that any one object is larger or smaller than another. A child of the same age once said to me : "You not naughty ; baby much not naughty," wishing to express that I was not naughty, and he was still less so than I. He delighted in using metaphors to excite laughter : "You are . . ." then he would pause to excite my curiosity, or to seek for an idea which was perhaps slow in coming ; "You are a *duck*, or else a *leaf*, an *acacia*, a *knife*, etc." These were absurdities, but they indicated **habitual** efforts to find out likenesses, **however** ridiculous they might be. When three years **old, wishing to** say something pleasant to a **little** girl he was **very fond** of, he said to her: "You **a** rat, a real rat, a pretty little rat!" Already, for several months, he had been in the habit of looking fixedly and very attentively **at** new faces, to take stock of them; and **after** having studied them for some moments he would say: "Beard like papa, blue **dress like mamma,** watch like grandpa. . ." At this age he knew the names of more **than twenty trees,** and could indicate the **most apparent specific characteristics** of each; this showed **great progress in the faculties of observation and comparison.**

## IV.

### GENERALIZATION.

It has been for **a** long **time admitted that** language is **a** necessary instrument for fixing, and indeed for forming, general ideas. It seems to me however that we may observe the rudiments of these **processes** in animals and children who cannot yet **speak.**

Animals **furnish us** with instances **of** an initial form of generalization. The **farm-dog,** who, **by** virtue both of his trade and his **taste,**—

> "Donne la chasse aux gens
> Portant bâton et mendiants,"

recognises **them in the first instance** by **their** dress and their *bâton.* These specific characters are engraved in the dog's intelligence, and associated with the mechanical necessity of barking. It is all **very well** for Bossuet and his disciples **to** urge, that if dogs bark at people of a certain aspect, **it is because their** masters **have taught them** this quite **mechanical habit.** Why then **do dogs not** confound one beggar with **another** beggar? **Some** beggars will excite them to fury, **while** others, on the **contrary,** have the power of gaining the dog's goodwill. **From a** distance, the dog will **bark at the** one as **well as the other**—this is the effect of the specific character; but on nearer sight he will soften down—this is the effect of individual character.

In all garrison towns one sees dogs who are particularly sympathetic towards soldiers; the sight of red trousers will make them run up to their owner. But when close to the soldier, they behave differently, according to the individual; they will caress one with eager joy, another they will treat with indifference, another perhaps with defiance. After the general idea awakened by the sight of the uniform, and which is equivalent to caresses, friendship and play, etc., there rise up remembrances of good or bad treatment associated with such and such a particular resemblance.

M. Houzeau is of this same opinion. "While Huber was making his splendid observation on bees, one of the hives met with an unforeseen accident. Part of the honeycomb became detached from the partition to which it was fixed, and slipping down several inches rested on the floor of the hive. It was not in the power of the bees to raise it up; its weight was too great for their physical strength. So they confined themselves to making it secure in its new position, by constructing fastenings of wax, with here and there abutments or props. But at the same time they thought of another necessary piece of work, that of consolidating the different combs which had not yet met with accidents. They fortified all the former points of attachment with fresh wax mixed with propolis." If this act of prudence does not attest the power of generalization, we should be glad to know under what name it should be designated. Bees are not in the habit of thus consolidating the fastenings of their honey-combs. They resolved on this precaution after the failure of their labours, and they extended it to all the combs which still remained in their proper position. They had evidently concluded from a particular case to the general.

Here is an analogous example in the case of mammals. "In October, 1859, I had made one day a long topographical examination of the watershed between the valleys of Rio-Frio and the Nucces. My animals had been without water from the time of my departure at four o'clock in the morning.

Towards three in the afternoon, I had finished certain sur-
veying operations ; and I mounted my horse and, descending the side of the little hill on the summit of which I had
spent part of the day, I took the direction which would
lead me soonest to a watercourse. After crossing a stretch
of undulating ground, we entered a large prairie almost
bare of trees, and which stretched for ten or twelve miles in
front of us. The ground was smooth, but furrowed at intervals by little tortuous trenches, a few inches in depth, hol-
lowed out by the rain in the wet season, but dry at all other
times.

"The land having no decided slope, these various furrows
did not unite in veins ; but ended in little ponds or minia-
ture lakes, which were scattered over the plain, 'like the
spots in the skin of a panther,' to borrow a comparison from
Strabo. But at the time when I was crossing, all the little
conduits were dried up, and there was not a drop of water
to be got from them. I had two dogs with me, who were
suffering cruelly from thirst. Scarcely did they perceive the
first of these furrows from a distance, than they scampered
towards it full gallop, descending unhesitatingly in the direction in which the water had once flowed. After a run of a
few hundred yards they reached the dried-up pond, and
after a short examination of its solid bottom returned to
me, evidently disappointed. The same thing went on during
the whole journey across the plain, which lasted till the
close of day. In this space of time the dogs explored between forty and fifty furrows. They invariably recognised
them from a long way off, rushed eagerly towards them, and
followed the dried-up course to its end. One could not
maintain in these circumstances, that the dogs were led to
these lakes by the smell, or by the effluvium of the water,
because there was not a drop to be found. They were not
guided by the character of the vegetation, for there was not
a single tree to be seen either along the furrows or by the
ponds. There was not even any particular kind of grass growing, so short a time does moisture last there. The dogs in
this case were guided by *general* ideas, seconded to a certain
degree by experience, and of a very simple kind no doubt ;

but in our conceptions of these furrows, of their origin, and of their use, they and I evidently reasoned in the same manner.

"I must add, moreover, that this observation is by no means an isolated one. I have simply chosen it as a particular case which is safe from various objections. I have many times seen, not only dogs, but horses, mules, oxen, and goats begin to look for water in places where they have never been before. They must have been guided by general principles, since they went to ponds or streams which were, at the time, entirely dried up."[1]

In default of words, the actions and gestures of children will often indicate to us what we are warranted in inferring concerning their power of generalization. For instance, a baby of eight months amuses himself for several hours a day seated on a carpet in the middle of the room. One of his favourite toys is a tin box which he likes because of its metallic sound, but especially because of its opening, into which it is his great delight to stuff anything that *will* go in or *won't* go in ! He has discovered that many of his little possessions, such as a pail, a little cart, a bottle, a trumpet, etc., etc., have this capacity for holding other things, and so, when any fresh toy is given to him, he instantly begins to examine it, to see if it has an opening.

The other day he got hold of a dressed doll ; and he tried to stuff a smaller doll, a piece of bread, and a little phial between its legs. Another time the stopper of a bottle was given to him, and because it was transparent he persisted in thinking there must be an opening at one end ; and he tried to put several things into this supposed hole. One of his favourite tricks is to put his forefinger in the eyes of people who hold him on their knees ; and this makes him laugh very much. In short, he has acquired a general idea of this quality of an opening and a capacity for holding things, which he has discovered in so many objects, and which he now seeks for in everything.

---

[1] Houzeau, *Etudes sur les Facultés Mentales des Animaux*, t. i., pp. 2 4, etc.

A child of eight months old always made starts of delight at the sight of any young or pretty person. Must there not have been in this case a distinction between a pleasant and an unpleasant appearance, which the child seized at first sight in all persons who came under his notice? This again was the germ of a general idea. The same child, when a fortnight older, manifested a desire for solid food, which he always recognised as such amongst many other things; whether it were bread, cheese, butter, fruit, meat, or sugar, he would lean towards it, stretching out his hand and making a vaguely articulate sound like "*wroua*"; he never did this when asking for his feeding-bottle or his mother's breast. At nine months old, the sight of a dog, a cat, a chicken, or a bird would send him into raptures; and he would hold out his arms, and look significantly at the person who was caressing the animal, as much as to say that he wanted to go close up to it. His movements for begging, and afterwards for showing his joy, always finished up with the repetition of the syllables *appa! appa! appa!* Thus he had a distinct idea, which, up to a certain point, was also a general one, of *solid food* and of *animal*, although he did not yet designate any of the individuals comprised in these two classes of objects by special names corresponding to the special qualities he did not fail to remark in them. At eleven months the exclamation expressing a general idea was changed into *ah!* and a few of his specific ideas were expressed by words of his own which need not be quoted here.

At the time when every one admits that children possess general ideas, that is, when they can express a good many of them by words, we can observe in them other general ideas which are not expressed by words. A little child of thirteen months knows very well how to say *matyé* (*marcher*) when he wants some one to take his hand and lead him about. But if his wish is refused, he knows how to fall back on his own resources; he leaves the person who is holding him, slips down on the ground, and crawls along on his stomach. Whether he wants to go straight along or to make *détours*, to go up stairs, or even down stairs (which he does less adroitly, and not without tumbles), crawling is his general means of getting along by himself.

If, when he is seated on the ground, the fancy suddenly seizes him to get up by leaning on his two hands, once upright on his legs, he is very often much embarrassed by his vertical position ; and he will entreat the first comer for a hand, whether it is a person he knows or not. Here again is a clear idea of a general means of getting out of his difficulty—to take hold of the hand of another person. This same child, who designates by the word *peau-peau* (*chapeau*) every species of headgear, bonnet, hat, night-cap, etc., etc., always puts on his head any object of the kind that is within his reach. There are also a certain number of things which he mistakes for this article of clothing, as for instance his nurse's basket, a paper-bag, a dish-cover, a lamp-shade, a handkerchief, and other objects, whose shape recalls more or less the idea of an object suitable for covering the head. This general idea has thus assumed extensive dimensions in his mind.

From all these facts, which each one can multiply for himself at pleasure, we infer the existence in quite young children of general ideas independent of language. When any particular characteristic has struck them vividly in a certain number of objects, it begins to fix itself in their intelligence in the shape of an abstraction, that is to say, of a very clear idea but no longer associated with the precise idea of certain objects. It is a sort of analogy summed up in a vivid conception which every object more or less similar is capable of reviving. When a means of fixing this conception is acquired, it assumes definite limits, either contracting or expanding, and becomes a true general conception. Words are only a more simple and efficacious means of recall than the pure sensation which was formerly wont to re-awaken this idea. Whatever may have been said about it, language is more an instrument for defining and fixing than for forming general ideas.

Let us observe a child generalizing after our example, but in his own fashion, with the words which we have taught him. To know things, is to distinguish the principal boundaries which separate them into different classes. To know a language, is not only to know words, but things, to have at one's finger's end, ready for daily use, all the observations, and experiences which language comprises. We can clearly see the part that words play in the working of thought, by noticing the irrepressible tendency of children, I do not say to classify, but at once to connect a known term with the various objects they see. With their unconscious and happy audacity they often give us most interesting examples of inferior generalization. Reason in children is, generally speaking, nothing more than comparison as little abstract as possible. The town of Tarbes has a pretty public garden which bears the name of the donor. A child from Bordeaux twenty months old, said to us: "There is a *jardin Massey* at Bordeaux." A child fifteen months old has a wooden horse christened *dada.* He only needed one single example, one single resemblance, to apply this word *dada* to one single horse (in which he was somewhat aided by his parents), and afterwards he would apply it instantly to every horse he saw. Here again there are only particular similarities, not conceptions that can be called general. In the middle of a court-yard filled with all sorts of animals, the child saw some chickens, which he called *koko,* from the name of his canary and turtle-dove; he saw some ducks and geese swimming in a pond, and he reduced them to one and the same species—the duck; a swan also is to him a duck. The broad lines of classification are already definitely acquired by the age of three; and a single point of resemblance suffices for the first rough outline, an enumeration (of instances) far from complete, for subsequently extending and confirming them.

There is, I think, some subtlety in the following mode of reasoning: "If it is true that the distinct general idea is posterior to confused particular ideas, conversely it may be true to say that the confused general idea is anterior to the distinct particular idea. Thus, the idea of man, in so far as it is characterized by the abstract and classical definition of a reasonable animal, or by the zoological definition, presupposes doubtless a comparison between many individual

204 THE FIRST THREE YEARS OF CHILDHOOD.

men ; but the confused sense of what there is in common
between all men is pre-existent to the definite distinction of
individuals : for instance, it takes a child some time to distinguish his father from other men ; it takes a dog some
time to distinguish his master.   It has been said with
reason, that the faculty of generalizing is a characteristic of
intelligence : it may also be said, that the faculty of indi-
vidualizing is no less essential a sign."[1]   I accept the
conclusion, but I make my reserve as to the premises.
When a child applies the word *papa* to all the individuals
who resemble his father generally, he only sees in them a
particular resemblance.   The author above quoted sees in
this an absolutely generalizing tendency, and he confirms
his opinion by adding : " One does not find that little
children generalize the word *mamma* like the word *papa*.
This no doubt is owing to the fact that, being more with
their mother than with their father, they individualize her
better."   Might not one reason for this difference (which
must not be exaggerated) be, the habit that the mother and
the women about a child have of making particular demon-
strations at every appearance of the father?   " See, there's
papa !  Where is papa?  Look at papa.  Say, ' How do you
do ' to papa," etc.   Thus instructed, the child sees papa
everywhere, and when he begins to speak brings out the
word *papa* at all moments, in season and out of season.
Moreover, are there no examples of children generalizing the
word "mamma" in quite as elementary a fashion as the word
"papa"?   Here is one, at any rate.   A child, three years
and five months old, had returned with his parents from
Tarbes, where he had spent more than a month.   At the
beginning of the journey, he asked every ten minutes where
were his aunts and his grandmother?   What were they
saying?   When would he go back to Tarbes?   Afterwards
he attracted the attention of a lady, whom he amused by
singing to her "St. Anthony," "Little Ida," "Marlborough,"
and the " Little Bird."   Then he became confidential towards her.   He told her that he had *five* mammas ; first of all

---

[1] P. Janet, *Traité Elémentaire de Philosophie*, p. 165.

his mother, then Aunt P., Aunt V., Grandmamma of Tarbes, and Grandmamma Louise; the mother of the latter, the great grandmother, was not a mamma. " But," said the lady to him, " Aunt V. is not your mamma ! " " Oh, yes; because she takes care of me." Then he began to imitate a grocer, his mother's grocer, the only one he had ever seen. Grandmamma Louise had given him a little grocer's shop. He said that he sold a great many goods, " M. François and Mme. Collette " (the heroes of a little story that had been told to the child) " often have colds, and then they come to buy the little black bonbons of me." Here everything is *particular*—the matter, the form, and even the incidents, which may vary according to the characters of the children, the vivacity of their imagination, the education they receive, and above all the examples set them.

We can now affirm that children do not begin, either by general conceptions or by terms which they transfer from the individual to the general, in virtue of that primitive tendency to generalization which, according to some philosophers is exercised before any individual discrimination. Max Müller has said, and M. Taine has repeated after him, that there are no general ideas without words. "There is in every language," says Max Müller, "a certain layer of words which may be called purely *emotional*. It is smaller or larger according to the genius and history of each nation, but it is never quite concealed by the later strata of rational speech. Most interjections, most imitative words, belong to this class. They are perfectly clear in their character and origin, and it could never be maintained that they rest on general concepts. But, if we deduct that inorganic stratum, all the rest of language, whether among ourselves or among the lowest barbarians, can be traced back to *roots*, and every one of these roots is the sign of a general concept. This is the most important discovery of the science of language. . . . These concepts are formed by what is called the faculty of abstraction, a very expressive term, which designates the action of decomposing sensuous intuitions into their constituent parts, of stripping each part of its momentary and concrete character.

. . . How is this work of the human intellect, the forming and handling of concepts, carried on? Are concepts possible, or, at least, are concepts ever realized, without some outward form or body? I say decidedly, No. If the science of language has proved anything, it has proved that conceptual or discursive thought can be carried on by words only. There is no thought without words, as little as there are words without thought!"[1]

M. Taine, who quotes this passage from Max Müller, adopting the opinions it expresses, has considerably enlarged on this interesting subject. In a chapter of his book on "Intelligence," where he has collected together observations of the same nature as those of Tiedemann, he speaks appreciatively, as a free disciple of Locke, of a certain number of facts relative to the formation of general ideas in children. He attributes the origin of general ideas simultaneously with general terms, or on the occasion of the latter, to an operation special to man, which he designates under the vague term of *tendency* to generalization. The examples which he has brought forward in support of his theory, will lead us to very different conclusions from his.

"The formation of these general names may be narrowly watched; with little children, we take them in the act. We name to them such and such a particular determined object; and, with an instinct of imitation common to them with monkeys and parrots, they repeat the name they have just heard. Up to this point they are but as monkeys and parrots; but here there appears a delicacy of impression which is special to man. We pronounce the word *papa* before a child in its cradle, at the same time pointing out his father. After a little, he in his turn lisps the word, and we imagine that he understands it in the same sense that we do, or that his father's presence only will recall the word. Not at all. When another person,—that is, one similar in appearance, with a long coat, a beard, and a loud voice,—enters the room, he calls him also *papa*. The name was an individual one ; he has made it general. In our case, it is applicable to one person only ; in his, to a class. In other words, a certain *tendency* corresponding to what there is in common to all persons in long coats, with beards and loud voices, is aroused in him in consequence of the experiences by which he has perceived them. This tendency is not what you were attempting to excite, it springs up spontaneously. In it we have the faculty of language. It is wholly founded on the consecutive tendencies which survive the experience of similar individuals, and correspond precisely to what they have in common.

We see these tendencies continually at work in children, and leading to results differing from ordinary language ; so that we are obliged to correct their spontaneous and too hasty attempts. A little girl, two years and a half old, had a blessed medal hung at her neck. She had been told, " C'est le bon Dieu," and she repeated, " C'est le bo Du," One day, on her uncle's knee, she took his eye-glass, and said, " C'est le bo Du de mon oncle." It is plain that she had involuntarily and naturally constructed a class of objects for which we have no name ; that of small round objects, with a handle, through which a hole is pierced ; and hung round the neck by a ribbon ; that a distinct tendency, corresponding to these four general characters, and which we do not experience, was formed and acting in her. A year afterwards, the same child, who was being asked the names of different parts of her face, said, after a little hesitation, on touching the eye-lids, " These are the eye-curtains." A little boy, a year old, had travelled a good deal by railway. The engine, with its hissing sound and smoke, and the great noise of the train, struck his attention ; and the first word he learnt to pronounce was *fafer* (*chemin de fer*). Henceforward, a steamboat, a coffee-pot with spirit-lamp—everything that hissed, or smoked, or made a noise, was a *fafer*. Another instrument to which children have a great objection (excuse the detail and the word—I mean an enema) had, naturally enough, made a strong impression on him. He had termed it, from its noise, a *zizi.* Till he was two-and-a-half years old, all long, hollow, slender objects,—a scissor-sheath, a cigar-tube, a trumpet,—were for him *zizi;* and he treated them all with distrust. These two reigning ideas, the *zizi* and the *fafer*, were two cardinal points of his intelligence, and from them he set out to comprehend and name other things." [1]

Let us call by their real name these *tendencies corresponding to what there is in common* between similar individuals or objects : they are embryo, if not complete, acts of generalization. There is a simpler and, to my thinking, a more accurate manner of interpreting the facts quoted above, as well as the analogous facts which I myself have been able to bring forward. Objects which are similar awaken the same ideas in the intelligence of children. At first, in consequence of their feeble capacity for analysis, these ideas of similarity are particular ; but by dint of frequently seeing similar things, simultaneously or successively, they perceive in them distinctive characteristics, or differences ; they cease to confound one with the other, they no longer take every gentleman with a beard for papa, all hissing objects for *fafer*, or every round object for *Bo-Du;* yet they nevertheless retain the idea of resemblance which they had grasped at first, and which is reawakened at every fresh sight of the objects, since, while now distinguishing them clearly from one another, they still call them all by the same name.

Here language is much behind thought : if the common term corresponds to the general conception, the particular or individual idea has not yet its equivalent in the child's vocabulary. A certain colour, a certain form, have suddenly made him aware of the presence of some sort of food, and he pronounces the general term by which he expresses the idea of *good to eat ;* but at the second glance, he has distinguished cake from bread, potatoes from butter ; thus from the general idea there have been detached particular ideas, which he does not know how to express. When he

---

[1] Taine, *L'Intelligence.* See Eng. Trans. by F. D. Haye, p. 15.

has set terms to express them by, the first general terms will
merge more and more into the particular, and he will invent
or acquire other terms by which to express the former
general ideas, which, on their part, will go on enlarging.

The capacity of children to generalize before they can
speak, seems to me to be precisely established by what
M. Taine calls this sudden tendency to generalize terms
which to us are individual. Such marvellous virtue in
words would be contrary to the law of intellectual evolution. If general ideas did not exist, in however rudimentary a degree, before the terms which are their cor-
relatives, I should see in the latter an effect without a
cause, the less producing the greater, the sign, the thing
signified. When you show a child his father, and say to
him, *papa*, you supply him with a word, which associates
itself in his mind with an already existing idea of a form
of such and such a kind. But this idea was not particular
in the child's mind; he had already seen this form before, and he sees it nearly every day; it is a vague idea
of resemblance, which is not abstract to him, but which
is re-awakened at the sight of all similar objects. The
words *papa*, *fafer*, etc., signify to him qualities which have
struck him, not in one object only, but in several. If the
child designates by the same name all similar objects after
having seen a certain number of them, and without taking
one for the other, it is incontestably shown that analogy
touches closely on generalization, and this for the
reason` that analogy of ideas tends towards a certain
generality. Words progress like ideas and by means of
ideas.

Generalization is, in fact, only more or less extended
similarity. It has not yet become (even when they begin
to exercise the function of speech easily) that superior
faculty of applying an abstract idea of quality to a whole
group of objects compared among themselves. The steps
which lead them to these distinct ideas of kind and species
will be very slow and gradual.

A child of three years old, of highly developed intelligence, could not at all understand words signifying species

or class. "What does that mean, this animal is of the same kind?" he persisted in asking me. I could only get out of the difficulty by answering: "It means, that the animal is almost the same." He understood that the one was like the other, and that was all. There is nothing, moreover, in which we find so much difference, even among adults, as in general ideas, especially the way in which they are understood. Take ten persons at hazard, and mention in their presence the terms virtue, humanity, force, law, nature, quantity, quality, or any other general term; ask each person the sense which he or she attaches to these words of every-day usage, and you will be astonished at the differences there will be in the answers; and the reason is, that the greater the power of thought, the more experience and analysis intervene in our intellectual operations, and the more general terms and ideas will define themselves, whether by contracting or expanding. *Bo-Du* will gradually cease to be the name of all round hanging objects, *papa* of all men with coats, beards, etc.; and in proportion as objects assume their individual names, or the names considered as such, they will take their place with the corresponding ideas in other general categories. The analogies first noticed become less striking, differences are more and more clearly perceived; classes are divided into species, species into varieties, and individualities into singularities. There is, as it were, a progressive acuteness in the intellectual vision, which at first grasps great masses only, afterwards the more important details, and finally the small minutiæ, and which corresponds to our generalizations successively ascending and descending.

## V.

### REASONING.

"The process of judgment has then for its special characteristic, according as it advances, the privilege of extending itself; of determining the reaction of the surrounding cerebral elements; of searching, to some extent, into the archives of the past; of associating former notions with

those of the present; of creating **partial local judgments,**
established *à priori* as results of the **inner experience of** the
individual; **and** of permitting us, at a given moment, **to**
juxtapose and agglomerate partial judgments—to agglutinate
them, in the form of arguments, into a complete judgment,
which resumes them all in a true synthesis." [1]

If then reasoning **consists** in the **fact** that the pre-
sentation **of certain phenomena** (which **already** have their
**equivalent in different** psychic **states** produced by past
**experiences) excites** these **psychic states** to reproduce them-
**selves, either wholly** or in part; in other words, if reasoning
**is nothing else than** a series of consecutive judgments co-
**ordinated** according to the law of habitual associations, it is
**evident that both** little children and animals **are able to**
**reason.**

A child **of seven** months has **already very well associated**
in its mind **the idea of the movements** of mastication with
that **of the agreeable sensations resulting** therefrom. When
he **sees** his **nurse lift any food to her mouth, and sees her**
lips and jaws **moving, he** judges that she is eating; **that**
what she is eating is **good to** her taste, **and** would be the
**same to his;** and **he knows** by experience that his nurse
would probably **let him share** this pleasure if he asked her
in an irresistible manner, *i.e.* if he cried, or pretended to be
going to cry—and **he acts accordingly.** We can see here
the origin **both of the analogical reasoning by which** he
**formed** this **chain of consecutive judgments, and** of the
**deductive reasoning which made him apply these** experi-
**ences which he had generalized to the** present circumstances.
**Before continuing the examination** of the faculty of reason-
**ing in little children,** let us study its **analogous** phenomena
**in animals.**

A **young dog, about six** months old, had been given to
me. **In order to form in** him habits of cleanliness, which
his former **master had** neglected to teach him, I used to
whip him every time that he made a mess in a room. Soon

---

[1] **Luys,** *Le Cerveau et ses Fonctions.* See Eng. Trans. (Internat.
**Scient.** Series), pp. 293, **294.**

the idea of punishment becoming associated with that of a particular need, excited in him the idea of acting in such a way as to prevent punishment. He took to waking me up every night, either by scratching vigorously at the door of my room, or, if this did not have the desired effect, by setting up a piteous howling. All the judgments which he performed in this instance were linked together by such close bonds that we have only to put them into words to see in them the elements of perfect reasoning. Let us try to utter his thoughts. " My master whips me soundly when I make my bedroom dirty (first inductive reasoning). But when, having opened the door, I go out for a little while into the court below, he is pleased with me, and instead of punishing me, he pats me and praises me (second induction). Now, when I make a loud howling, he wakes up and opens the door for me (third induction). Come then, I'll bark with all my might, wake him up, and I shall not be beaten." (Deductive reasoning.)

The following fact is not less conclusive. I borrow it from the immense store of observations and quotations of M. Houzeau, who was a witness of the incident. "It is known that the milk-sellers of Brussels employ dogs, harnessed to little carts, to go every morning the round of their customers. These dogs draw up of their own accord at the different houses ; but in this there is nothing more than memory. One day in 1854, I was walking along the Rue Saint-Géry, and I happened to follow one of these dogs drawing its cart, while some steps behind him followed the milkwoman. A carriage with two horses was also going along the street, at the same pace, side by side with the milk-cart. This carriage formed a constant obstacle on the dog's left hand, as it was between him and the line of houses, one of which he had to stop at. The problem in his mind was, whether to cross in front of the horses, or to let them pass on and cross over behind them. The dog went on walking at his usual pace, casting alternate despairing glances at his mistress and at the door of his customer. The expression of the animal's face said unmistakably : ' What shall I do now ? ' The interrogation was so plain and positive that the milkwoman not only understood and answered it, but she solved the problem in the manner that the dog's look seemed to suggest. She begged the coachman to stop his horses for a minute, and the dog then instantly crossed over in front of them and drew up at the right door. Any one who has witnessed a similar scene cannot deny that dogs are capable of reflection."[1]

"At the time of the great inundation of the Loire, in 1836, the water spread over a garden where two nightingales had built a nest in the hedge. The floods mounted higher and higher, and threatened to submerge the young family; for the little ones, but newly-hatched, were not yet capable of flying. Under these circumstances, we might assert that the old birds must have been able to reason in order to understand the increasing danger. But, unquestionably, there was something more than automatic action when the birds carried the nest bodily off, and placed it at some distance, out of reach of the waters. This is, in fact, what the parent birds did; each took one side of the nest by the beak, and with a swift and balanced flight the nightingales accomplished the journey and saved their little ones from the threatened drowning."

I will quote another example of a different kind, also about birds. "I had received a present of a beautiful male grossbeak," writes Audubon; "but the bird was in such an exhausted state that one would have said he was simply an inanimate mass of feathers. However, with careful feeding, he soon recovered, and became so tame that he would eat out of my hand without showing the least sign of fear. To make captivity bearable to him, I allowed him to fly about my bed-room, and on getting up in the morning, my first care always was to give him some seed. It happened, however, that for three consecutive days I lay in bed later than usual, and then the bird came to wake me by fluttering on to my shoulder and demanding his usual meal. The third day I let him flutter about some time before appearing

---

[1] Houzeau, *Facultés Mentales des Animaux*, vol. ii., p. 195.

to be awake. But he no sooner saw that he had attained
his end, than he retired to the window and waited patiently
till I had got up."

Thus we perceive in animals a variety of intellectual
operations and intelligent actions which cannot be set down
to instinct. The examples above quoted do not belong to
the class of habits common to a whole species, which we
might consider as dependent on the organic constitution,
but they are individual manifestations, under exceptional
circumstances, and varying with external changes. It is
human reasoning, with all its independence of automatism.
We must, however, be careful, whether as regards human
beings or animals, not to attribute too much to spontaneity
pure and simple. Are we not constantly startled by the
sudden apparition of some faculty long buried in the depths
of hereditary automatism? and may not certain exceptional
actions performed by animals proceed from the same mysterious source? Are inundations, for instance, such rare
events in the life of the species, that they may not have
determined in birds up to a certain point the semi-instinctive
faculty of transporting their nests by the help of their beaks?
In like manner, in the case of the child above mentioned,
who watched his nurse eating with envious eyes, there must
have been combined with the conscious operations of the
intellect, with the process of pure reasoning, certain reflex
judgments and movements which were the result either of
individual or of transmitted habits. Might not the mere
sight of the movements of the jaws excite in a child unconscious and involuntary movements, such as opening the
mouth, holding out the arms, leaning forward, and even
crying? It is very difficult to determine how much belongs
to unconscious and how much to conscious cerebration, in
this aggregate of apparently rationally co-ordinated organic
sentiments, ideas, and impulses. It is certain, however,
that consciousness enters largely into the matter. In fact,
as we have already said, all instinctive movements, all
reflex actions, come in course of time to be consciously
realized by their agent, provided they are important enough
in themselves, or that by quickly recurring reiteration they

are brought out, so to say, in relief. In adults we see the consciousness suddenly roused by great crises or theatrical climaxes, whereas ordinary events leave them indifferent. In like manner, if children are encouraged to reproduce these movements, which at first have passed unnoticed by them, or to perform them with varying or increased intensity or complexity, their attention is sure to be attracted sooner or later, as is proved by the fact that they will modify these actions under the counteracting influence of particular impressions or sentiments.

A little child of ten months wished very much to hold in his arms a kitten which his elder sister was playing with on her lap ; he held out his arms towards the animal, looking alternately at it and at his sister, and uttering cries which sounded like hiccoughing. His sister did not grant his desire, not wishing to subject her pet kitten to the awkward hugging of the baby. The latter then began to scream. His sister remained calm and impassive. The baby then became more urgent, shook his whole body violently, struggled, writhed, sobbed, and howled. His sister said to him : "Be quiet, you naughty child, you shall not have pussy, you will hurt him ;" but the child's passion only increased in fury, and finally reached such a pitch of intensity that his consciousness quite disappeared in the hurly-burly of moral and physical excitement, and he ended by forgetting the cause of his rage in the rage itself. He went on screaming and holding out his arms without knowing what he was doing, for the kitten, terrified by this turbulent scene, had escaped and hidden behind a bed. To calm her little brother, the sister called the kitten back and took it to him ; he instantly became quiet, but after a few seconds he began to cry again ; the sister then kissed and petted him, and tried to make him stroke the kitten. When at last he was a little tranquillized, he took no more notice of the kitten, but asked his sister for a bonbon. Here we see the fluctuating and capricious procedure of a child's judgments and movements, which, though co-ordinated by the law of habit, are shifted and changed in a thousand unexpected ways at the beck of intervening impressions and associations of ideas

which alternately present them to and withdraw them from
the consciousness.

If automatism **passes** back **to** consciousness in certain
determined cases, consciousness is equally ready to relapse
into **automatism.** A child of two years was accustomed
**to see his** mother's lodger (Mme. Jillet) re-enter the house
nearly every day on her return from market. **Whenever he
saw** his mother come back with a basket on her arm **he**
used to call out, "Where mé Gilé? Me want see Gilé.
Call mé Gilé." Are not all the acts of judgment expressed
by these baby phrases linked together in such a way as to
constitute a **chain of concrete** but effectual **reasonings?**
Well then, **the child has** already **acquired the habit of**
repeating these phrases **several** times **a day, and no matter**
at what hour, and quite mechanically, **like a parrot, every**
time he sees his mother take up a basket. What was
rational has become automatic, **by dint of constant**
repetition. I observed other examples of the same nature in
this child. When the lady in question comes to pay her
morning visit after marketing, the child runs up to her,
hangs on to her gown, and says: " Maman dear Gilé," then
he pulls open a corner of the basket, forages about with
his hands among the provisions, and pulling out no matter
what, says, "Want this, I do ; I say, I want this." All
these natural **operations of judgment** happen in so constant
an automatic order, that **one would** suppose **that the** child
had no consciousness of them.

But here are two examples of a contrary nature, furnished by the same child, from which we shall see the
**intervention** of consciousness in a collection of judgments and movements performed **chiefly in a reflex** manner.
His father is a workman **who often goes out fishing, and
the** child is accustomed to eating **fried fish.** The other
day, the father, who had returned home after the **hour** for
the family supper, was eating some fish by himself. " Me
want fish, papa, me want fish." The father turned a deaf
ear for a few moments. " Me **want** fish," again said the
little one, pulling his father's sleeve and trying to attract
his attention. The father went on eating without speaking

a word (up to this point all the child's words, as well as its actions and gestures, had expressed ideas automatically associated and co-ordinated) ; but finding his efforts fruitless, he suddenly got under the table, and pulling his father's leg began again : " Me want fish, not pussy have fish, me want fish." He had suddenly reflected that the cat generally went under the table ; and by a sudden inspiration of his consciousness, which made him imagine and execute movements quite new to him, he assumed the character of a cat.

The mother of this child has accustomed him to call Madame Jillet, towards twelve o'clock, and to ask her to throw down the newspaper, in order to save her the trouble of coming downstairs. The child discharges this duty very solemnly, and will not let any one else do it for him, " Mé Gilé, give paper !" He repeats this formula with very little variation until the paper is thrown out of the window. One day his brother wanted to call out for the paper instead of him. The child became crimson with rage; his rights were being invaded. He set off screaming with all his might : " Mé Gilé, give paper !" and then, turning to his brother ; " You not to call out, musn't call out, me cry out, mé Gilé !" The paper fell at this moment, and the elder boy picked it up ; the younger one instantly snatched it from him, saying : " Give paper, me take mamma." Two or three innovations had here crept into his accustomed routine ; but he returned to this immediately, shaking the paper which had fallen, and saying, " Not dirty ;" this had become an almost unconscious habit with him.

It is in this faculty of assimilating past experiences with new ones, of continually extending the chain of inductions and deductions, that children and animals show endless power of invention and resource of imagination—a power which adults, more accustomed to act on ready-made arguments, acquired or learnt, than to construct new ones, might envy them. *A propos* of this example, we may observe that Locke and Mill have with reason maintained that there is a mode of reasoning founded on the particular.

M. L. Ferri, in his interesting study on *Les Trois Pre-*

*mières Années d'une Enfant,*[1] has introduced some observations which confirm this opinion.

Young children are incessantly giving proofs of their force of invention, and of the elasticity of their reasoning power. All their intellectual, moral, and physical progress, their games, their caresses, their little tricks, all bear the mark of the practical and ingenious nature of their reasoning powers. We will quote a few more instances selected out of thousands, and of the kind which every one may have observed, without however noticing or appreciating them systematically, and with a view to the right direction to be given to infant faculties. The word *direction* does not rightly express my meaning. Everything bursts forth so spontaneously in the evolution of young human beings, the experiences they are continually accumulating respond with such marvellous activity and diversity to the excitations and necessities of accidental circumstances, that the great art of education, even at the most tender age, seems to me to consist rather in attentive and watchful neutrality than in partial and domineering interference. Woe to the child cast in the mould of conventional routine and maxims, however wise and specious a form they may assume! "Let be," "Let pass," "Let live," and do not force or repress, except when absolutely necessary, or you run the risk of hindering the diffusion of the sap destined to produce, successively and simultaneously, precious flowers and exquisite fruit.

Here are some more examples from the category of children who can already speak and walk, and thus reveal in a more evident manner the workings of the unseen phenomena which are going on within them. The child I am about to speak of is two years and one month old. When his father, whose night duties oblige him to sleep rather late in the morning, seems to him to have slept too long, he tries to wake him up, if left alone with him for a moment, either by pulling off the counterpane or making a noise with a chair, or by getting up on the chair

---

[1] In the *Filosophia delle Scuole Italiane*, Oct. 1879.

and shaking his father's **head**. The other day, mounted on the chair, he **took** his father **by the** neck, and then **pinched** and pulled his ear. The father pretended to be asleep, **in** order to see what would happen. At last the child lifted up one of his father's **eyelids** and cried out, " Daddy, isn't it light?" This was **a device** that the father **would** have to guard against in future. Another time **the father** had just **come** in, **and was waiting** for supper; **the child** was watching **his** mother's **preparations**; she had just filled a plate with soup and **left the room for** a moment. The child profited by her **absence to take up the plate in both hands**; and, notwithstanding **its weight, he carried it to his father**, saying to him : " Monsieur papa, **mangez soupe**." This action was altogether new, and **caused a** burst **of laughter** from both parents. The *Monsieur papa* was **the most** comical part of the entertainment. **Sometimes**, when his father pretends to be angry, **the boy soon guesses, from his expression, and the** curve of his **lip, and other** familiar **signs, that the anger is** only feigned, **and he calls out, laughing, "** Not angry, papa play, **not** angry." **This child has taken to a trick** of slily stealing something from **his** neighbour's **plate at** meals. When he has transferred the bit of stolen meat or pudding to his own plate, **he pats and presses** it up against the rest, so that it should **not** be seen. This little trick has already procured him several scoldings, **which have** not yet quite **cured him of it.** He knows that **his brother, who** is **past five, and goes to school, does not** like **him to touch his playthings, because he disarranges and spoils** them. **But, no sooner is his brother** gone, than **he** goes off to the **corner** where the **box of** toys is hidden, or manages by **coaxing and** tears to get them from his indulgent mother. **But he is on the** *qui vive* **all the** time, and often runs to the top **of the staircase to see who** is coming up; **and the** instant he **recognises his** brother's step he begins with both hands to **bundle all** the **toys** pell-mell into the box, and to put them **back** in their place. Although it is indisputable that in many **cases** adult animals reason more promptly and **justly** than do **children** from **one** to four years, or even **older,** this is assuredly an example of contrivance superior to that which is ordinarily shown in the most intelligent animals.

As children grow in strength and experience, their judgment gains in accuracy, their reason becomes stronger, more precise, and more subtle,—more abstract, so to say,—and the verbal expression of their trains of reasoning, which relate generally to their desires and fears, acquires fluency and logic.

When I was a child of five years I had abused my right of strongest, and taken away from a little girl cousin, two years younger, a magnificent pear which our grandmother had given her. The tears and screams and despair of the poor little victim may be imagined. She related what had happened very eloquently and circumstantially to my grandmother, and the latter tried to appease her by giving her a fine bunch of grapes. When I came out of the hiding-place where I had been enjoying my stolen fruit, the little girl had not yet come to the end of her grapes or of her grief. My grandmother scolded me very severely, and then addressing my cousin: "Never mind, my child," she said, "I will fetch you a very nice pear, and your cousin shall not have one." This promise, and my disconcerted air, restored my cousin's cheerfulness, and she ran off merrily to her play. An hour passed by, and I had not left the house; probably I was waiting to see whether our grandmother would have the courage to fulfil her threat, and to give my cousin a pear in my presence without giving me one. The little girl, on her part, had not forgotten the promise either; she kept coming back every ten minutes, fidgeting round the old lady, and saying pretty coaxing things, asking if she could do anything for her, talking to her about flowers, vegetables, etc., but the word *pear* never passed her lips. She knew by experience that our grandmother did not like importunity, especially in regard to things to eat, and she avoided displeasing her by appearing too eager for the promised fruit. At last, however, at the end of half an hour, she screwed up all her courage, and nerving herself for all risks, came jumping into the room from the court-yard, and laying a coaxing hand on her grandmother's shoulder, said : " It will be very nice, won't it ? " My grandmother understood very well what this astute remark meant, but purposely made her repeat it ; and the child, thus encouraged, went even further : " It will be nice, grandmamma, the pear ! " My grandmother was charmed with this proceeding, and calling to my grandfather, who was in an adjoining room, asked him if he had time to go into the garden. " Will you go and look for a fine pear which I have promised this little girl ? " My grandfather went into the garden, and my cousin accompanied him, holding his hand ; I too made one of the party, but keeping at a respectful distance behind my grandfather. My cousin had a beautiful pear given her ; and she was so much engrossed with the pleasure of eating it, that she forgot to say that I was not to have one. So we all three enjoyed our pears together, and then we had a good game among the trees in the garden with our grandfather—two little children and one big one !

Children are thus capable of reasoning from their very cradles ; if it were not so, they would never acquire the power : according to the saying of Laromiguière : " Reduced to pure sensations, which they could neither disentangle nor compare nor combine nor analyse, they would be destitute of all ideas, and would never take rank among intelligences."[1] But the natural logic of young children is very limited in its grasp and very uncertain in its methods. If in the sphere of analogical relations it goes straight to ·the end, as if by an imaginative glance ; if in the rather more complex relations of particular similitudes, which lead towards induction, it follows the course, sometimes tolerably regular, of psychic associations ; and if it applies somewhat resolutely some of these analogical and inductive conceptions to new cases, it is the primitive, summary, and irreflective method which suits children best. This method is confounded with the most elementary acts of discrimination and classification ; often it is only a similitude applied to a passing fact, the most simple adaptation of movements to

---

[1] *Discours sur la Langue du Raisonnement*, p. 185.

representations. Many of the above examples are proofs of this. But it is a satisfaction to me to see my own interpretations confirmed by observations analogous to mine. Darwin, although he does not fix the positive beginning of association of ideas in children before the age of five months, nevertheless notices a sign of practical reasoning in his little son at the age of a hundred days, when he slipped his hand along the finger held out to him in order to introduce it into his mouth.[1] A child who cannot yet speak, but is beginning to walk, stumbles in passing from the bare floor to the carpet. A moment comes when it occurs to him to raise his foot to a sufficient height to step over the border of the carpet without stumbling. Here is a middle term inserted between the end conceived and the point of departure.[2] A little girl nineteen months old wanted to have my hat, which was placed on a table too high for her to reach : impatience, screams, tears. I get up and give it to her. Her first impulse is to put it on her head ; then, after a few minutes, during which she seemed to be thinking, she went to fetch her own hat, which was on a chair within her reach, and presented it to me with a most serious expression. Was it a mode of thanking me, or an invitation to me to take her out? It matters little ; but it was evidently analogical or even inductive reasoning.[3]

These are the modes of reasoning in which children excel, even after they have begun to talk. They seem less sure of themselves in those which they attempt at our dictation, on our system, and by means of the analogies and inductions which are imposed on them by our language. They judge, often very wrongly, from simple similitudes and forced analogies. "If they dislike a person very much, every one else who at all resembles that person will be repugnant to them ; they would readily lay down this axiom : All persons

---

[1] *Biographical Sketch of a Little Child. Scientific Review*, July, 1877.
[2] See the *Revue Philosophique*, April, 1880, on the above-cited paper, by L. Ferri.
[3] Paul Rousselet, *Pédagogie à l'Usage de l'Enseignement Primaire*, p. 246.

of such-and-such an aspect are bad. A premature and un-scientific induction, yet nevertheless, at bottom, of the same nature as those of *savants*." [1] It is not indeed very certain whether most of the trains of reasoning that children prefer and succeed in best from two to three years old, generally surpass the average reasoning of adult animals. I had a cat which, from the time she was a year old, was in the habit of taking her dinner with me. The instant dinner was ready she used to begin frisking about, from me to the table and from the table back to me, and try by persuasive looks to make me come to it. Do we find any superior element in the reasoning attributed by M. Egger to his son? [2] "From the age of eighteen months, as soon as meals were ready he used to summon all the absent members of the family and drag them by their clothes to join the meal. . . . At twenty-eight months I observed the scope of his reasoning powers extend. He now guessed that any person who took up his hat was preparing to go out, that in going out he would pass through the court-yard, and that he could see him pass there; the instant any one takes up his hat, he goes to the window to see him go." There is nothing in this, whatever M. Egger may think, which indicates any superiority of infant over animal intelligence.

The faculty of reasoning is always very limited in its logical scope during the whole three years of which we are treating. If language brings a child new elements of in-duction, new facilities, and especially, under the example of his elders, the to him new faculty of argumentation, it most often only serves to bring out in relief his intellectual feebleness; and a very useful result this is, both for us and for the child—for us, who are thus enabled better to con-trol the child's progress, and for the child, who has thus the opportunity to take stock of himself, and to realize the limits and the scope of his budding intelligence. No one

---

[1] Henri Marion, *Leçons de Psychologie appliquée à l'Education*, p. 340.
[2] *Observations et Réflexions sur le Développement de l'Intelligence et du Langage chez les Enfants.*

has better demonstrated than M. Egger these characteristic
facts of intellectual weakness in children who so often
astonish us by the precocity of their reason and the sagacity
of their judgment. The instances which we borrow from
him have moreover a double interest, psychological and
moral observation being closely bound up with the lexico-
logical considerations which they embody.

"The younger sister of Emile said to me: 'I shall carry
Emile when he gets little.' She had noticed that, of two
people, the biggest only can carry the smallest; she has
also been told that she will become big . . . the con-
ditional and the future are confused together in her mind."
This observation is of real value from all points of view,
and I can say as much for the two following: "The reversing
of relations is very habitual with children at this period of
life. At four years, and even at five years of age, a child
will take correlative ideas one for the other, like that of
*lending* and *borrowing.* 'Will you borrow your seal to
me?' He said to me one day: 'I am very generous
to-day,' meaning to say: 'You have been very generous to
me.'" We must however add that these infantine confu-
sions take place rather in words than in things. Let us
study the natural limits and the most common deficiencies
of infant reasoning in circumstances almost independent
of the admixture of speech.

## VI.

### THE ERRORS AND ILLUSIONS OF CHILDREN.

Children are subject to the same kinds of error and
illusion as adults; the only difference between them is, that
particular kinds of error are more natural and unavoidable
in the former, owing to the imperfect development of their
faculties, which nevertheless from the beginning of life tend
to exercise themselves according to the specific laws of
human thought. Apart from this difference, we find in
children all the most common errors, false inductions
applied to the impressions of the various senses, the para-

logisms occasioned by precipitation and prejudice, and above all, what we still call, after the logicians of Port Royal, sophisms of *amour-propre*, of interest, and of passion. To point out some few of these different errors, often analogous in children and adults, will not be a work of idle curiosity— it will throw a flood of light on the operations of the infant mind and afford us knowledge as useful for us who live in the life of our children, as for the children whom we have to prepare for life.

Plato has indicated very poetically and accurately the principal source of our errors, viz. ignorance. "Let us suppose our soul to be a sort of dove-cot for birds of every species, some are gregarious, others that consort in small numbers, some that fly by themselves through all the others, this way or that as it may happen. By these birds we must conceive that different kinds of knowledge are meant. While we are children the dovecot is empty; but every time that we acquire a piece of knowledge we let it loose in the dovecot. When we want to 'recapture' one of these pieces of knowledge, it may often happen that we get hold of one by mistake instead of the other; as for instance, if we were seeking the total of seven and five, we might get hold of eleven instead of twelve—our ring-dove, as it were, instead of our rock pigeon. But in this case it would be knowledge that would make us ignorant."[1] However subtle these last remarks may be, we may accept their imagery and profound meaning. It is, in fact, on the strength of a little knowledge that we judge rightly or wrongly; and the reasonings, whether conscious or intuitive, which lead us to error and to false judgments apply to the simplest phenomena of the mind and even to our perceptions. We shall find numerous examples of the truth of this amongst children.

Let us first examine the errors which the exercise of our senses gives rise to. The illusions of sight have to do with the colour, the localization in space, the form, the dimensions, the distance, the nature, the number of all objects which are

---

[1] Theætetus.

illumined by light. The sensation of light produces the notion or judgment of colour; the reality appears to us under colours with which we ourselves invest it by virtue of anterior judgments. What a time it must take a young child to accustom itself to recognise the colours belonging to the different surrounding objects, susceptible as they are of a thousand variations and transformations, according to their reciprocal actions on each other, the action on them of different surroundings, and according to their distance and position. A child of six months will gaze fixedly at, and then stretch out his hands entreatingly towards, the flattened sphere of a door-handle which he mistakes for the suspended ball with which his nurse has amused him in his cradle. He will take a flat disk with gradations of light and shade for a globe of uniform colour; as, later on, he will mistake a sphere for a square. At this age, however, his most frequent error is to take all the surfaces he sees for bulks. He wants to hold in his hands everything of which the brightness attracts him; and the reason of this is, that, owing to his limited knowledge of perspective, he does not judge of the distance of far-off objects in comparison with the relative distances of the objects which surround him: his scale of height and length is entirely circumscribed by the environs of his own person and the few places well known to him, like his room or some parts of his room.

As early as ten months, or a year, it seems to us as if children were enchanted by the solemn appearance of the sun or moon, when setting or rising; this however is an illusion on our part. A child who is not in the habit of contemplating these luminaries at other times, does not invest them at rising and setting with the voluminous appearance which we see in them on these occasions. I have even remarked that a child's attention is scarcely attracted by the brightness (much more relative than real) of the stars: their scintillation, to eyes which only see them as near objects, is very trifling. Children are constantly making strange mistakes respecting the height, the distance, and the form of numbers of things. Mountains, forests, horizons, which we see and reckon to be eight or ten miles

from us, to a child even of fifteen months old, will seem no
larger or more distant than such-and-such a neighbouring
tree or house. Far-off objects seem to him as smaller
objects near by, while near objects he almost invariably
thinks to be much larger than their real size.

A little later on, when a child is four or five years old, for
instance, his sense of perspective, being now a little more
developed, will lead him into quite opposite errors. I
remember that when I was five or six years old, I accompanied my mother and an uncle to Bagnères-de-Bigorre.
We had been for a walk on Mont Bédat, whose base is
planted with trees, shading winding paths. My uncle,
having left us for a moment, disappeared amongst the trees,
and we did not see him till he reappeared on a footpath a
little higher up. I thought he was an immense way off, but
I have since verified that he could not have been more than
a hundred yards distant. It is in consequence of psychical
(and not optical) errors that young children judge very falsely
of real forms, of movement, and of the distance of objects,
when the object is some little way off. If they are moving
along in a carriage or a boat, they go on for a long time
believing that the trees and houses they pass are moving;
when they see a long avenue, they think that the trees join
and touch each other. If two buildings at different distances off appear to them of the same height, they think
they are at the same distance, and of the same height. As
for rather distant objects, they are not capable of estimating
their real size by comparing their apparent size with the
distance at which they are. The utmost they can do, even
at two years old, is to estimate the distance, dimension, and
movement of near objects, within the reach of eye and
hand. But it has been noticed that the more intelligent
children are, and the more they are accustomed to games
and exercises which call forth strength, skill, and quickness
of sight, the greater is their certainty of visual appreciation.

The most frequent illusions of sight,—normal illusions *i.e.*,
not pathological ones,—are those which relate to the relief
of objects, to the confusion of what is visible with what can

be held in the hand, and to the number of objects within the field of vision. Children refuse for a long time to believe in the solidity of objects. They feel them all round and all over to find some cavity ; and they try to put all sorts of things one inside the other. A child of fifteen months who was very fond of hoisting himself up to the second landing of the staircase at home, on another staircase stopped at the first landing because the landing was so large that he thought the staircase ended there. There is no doubt that children, even after a certain number of experiences, will be very puzzle-pated in making distinctions, which we determine at first sight, between such things as the ledge of a chimney-piece, a roof, a portion of a wall, the facing of a bridge, and the plane surfaces which are contiguous to them. We can also well understand that a child, even of three years old, and who has some idea of concrete numerical quantities, will yet often make mistakes in estimating the number of objects presented to him. Is it not the same with adults? We know that to each one of us when we look up at the sky on a fine starry night the number of stars we seem to see is immensely exaggerated. Here we have an illusion entirely mental. An illusion of this kind takes place in a child's mind when it wishes to count a number of objects close to each other, even when but little removed from himself. At the age of three years he will perhaps be able to count, up to four or five, the trees which are quite near him ; beyond that limit the number becomes many trees, and the number increases with the distance of the objects. That which is true of numeric magnitude is also true of magnitude of extent. A pool of water seems like an ocean to an infant of two or three, and the sea is *much, very much water ;* but this vague idea of the indefinite is much more limited in children than in adults.

The sense of sight, like that of touch, must be exercised in recognising the intrinsic properties of objects ; in other words, the intuitive appearance only becomes an object of real cognition when it is referred to a genus which determines the object. When a child sees his nurse, at a few yards distance from his cradle, take up her bonnet and cloak, fill his bottle with milk, stroke the cat or dog, etc., these are forms of *thought* rather than visual perceptions. At the sight of these objects, or, let us rather say, on the evidence of his sight which brings to him some of the sensations which cause him habitually to infer these objects, the child affirms to himself—believes that he sees them. By virtue of judgments and reasonings as rapid as they are unconscious, the child immediately refers these tangible qualities to a known genus ; this is why the number of his errors, relatively to this specific appreciation of objects, is in inverse ratio to his accurate experiences and his well-drawn inductions.

Finally, we must not forget those hallucinations, in some sort legitimate, which are produced in a child's mind by undecided combinations of light and shade, a sudden passage from light to darkness, the juxtaposition of certain colours. These illusions are more optical than mental ; but from the tendency which all animals have to clothe in an objective form any sensations that are rather vivid, children,—and sometimes, too, adults,—are led to create chimerical spectres, real phantoms, out of these optical spectres. Nothing is more hurtful than to leave children alone in badly-lighted rooms, or to let them pass suddenly and alone from light into darkness. Some children, from a false conception of the relations between light and darkness —especially if the light flickers at all—have been the victims of terrors resulting from illusions which have taken bodily shape. We should try to accustom them gradually to the changing effects of light and shadow, and teach them to produce for themselves strange effects which they will no longer be frightened at when they have become accustomed to produce them themselves and to laugh at them. Thus a child of two years used to beg his mother to leave the candle till he was asleep, so that he might make dogs, elephants, and jaguars with the shadow of his fingers on the wall.

The sensations of hearing give rise to special errors, but

not such numerous or important ones as those caused by
visual sensations. The great difficulty for children, is not
so much to localize sound in space, as to refer it to its right
cause ; but these two cases seem often confounded in one.
As it is not essentially useful for a child to make this localization in the absence of impressions of sight, he does not
acquire the power till rather late, and after he has had a
great many combined experiences of hearing, sight, and
touch, and that his faculty of comparison has developed to
some extent. Thus we see a child of a year old very easily
mistaken as to the distance of a sound otherwise familiar.
Nurses and other people who are about children will often
amuse themselves by raising or lowering their voice, remaining all the while hidden, to make the children think
they are either quite near or far off. As to the errors
relating to the fusion in one whole of various sensations,
either synchronic or successive, whatever importance they
may have as regards the musical education of little children,
they belong to the chapter of infant æsthetics, and we
abstain from alluding to them here. Suffice it to say, that
the spiritual ear, or the faculty of judging sonorous sounds,
plays many tricks, even if it renders good service, to the
material ear.

It is difficult to say how much harm is done to children
by people who do not scruple to give vent violently before
them to their tempers and passions. Children generally
imagine that this or that angry tone of voice means displeasure with *them,* and that they are going to be scolded
and punished. It is the same with dogs and cats whose
masters lose their temper about every trifle ; when they
begin to storm or to swear, the animals hide themselves
in terror. I knew a little girl of five months who used to
be constantly whipped and scolded to cure her of a dirty
habit ; and often the mere sight of an angry face, an angry
gesture or voice, would be enough, from some mysterious
association of ideas, to make her commit the fault in spite
of herself. Another little thing of fifteen months, whose
father used to scold her unreasonably often, whenever she

heard a heavy step on the stairs thought it was her father, and ran trembling to her mother.

Smell and taste are liable to numberless causes of error. In the first place we know that the nervous excitability of the sense of smell very quickly wears off, so that we only enjoy pleasant odours or suffer from bad ones at the moment and during short periods. For each scent in its turn to be fully appreciated, they must succeed each other in a certain order. We know also that smell and taste are closely connected, and that they deceive conjointly. We can tell the taste of cinnamon from its smell, and if we pinch our nose while eating it, we find no taste in it but that of fir-wood. Many other substances lose their flavours in the same way. Thus, nurses act scientifically without knowing it, when they pinch the babies' noses to make them swallow disagreeable medicine. Taste deceives us in a similar manner. When we eat or drink substances of different flavours in quick succession, without giving the nerves of taste time to rest from their first sensations, however different the flavours may be, we are no longer able to distinguish them. By drinking alternately, with our eyes bandaged, buttermilk and Bordeaux wine, it becomes impossible after several repetitions of the experiment, to discover any difference between the two liquids. These are illusions which children's doctors do not fail to avail themselves of.

Finally, as to errors of *touch*. They are generally easily corrected by the attentive use of the tactilo-muscular organs. This is at any rate the case with regard to impressions of surfaces, their smoothness or roughness, hollowness or relief, etc. The exercise of our organs pure and simple corrects more or less quickly and completely the errors of the child in this respect. Needless to say that the intelligence of the child and the intelligence of the educator contribute much to hastening on progress of this sort. It is, moreover, often difficult, even for adults, to rectify the vague or determinate indications of the sense, or the senses, of which we are treating.

The appreciation of different weights is not an easy thing. If we take up two things together, we cannot always say which of the two is the heavier; or, if the weights are almost equal, we cannot determine whether they are exactly equal or not. This explains to us the confusion made by a three-months-old child, between a full and an empty bottle. It also explains to us why a rather more experienced child either attacks all objects alike and tries to lift them, or else hesitates to touch even the lightest objects, fancying they must be heavy.

The sense of touch is also subject to frequent errors with regard to appreciating the different degrees of heat in different bodies. But here children are scarcely more at fault than adults. Doctors tell us that the heat which exists in bodies in a latent state does not affect us. The temperature of our bodies remaining the same, as well as that of external objects, different objects affect us thermally in an unequal manner. In a room where all the objects are at a temperature of 120 degrees, we feel a lower degree of warmth if we put our naked foot on the carpet than if we touch the marble chimney-piece or any metallic object. At the same degree of temperature the walls and the wood-work appear much hotter than the clothes we have on. If we suppose the same objects to be all of them at a much lower temperature, the relative degrees will be reversed, and the objects will appear colder and colder, in the same order in which before they appeared to be warmer and warmer. These singular phenomena are explained in the first instance by the greater or less facility which different bodies have of communicating heat, and of producing on us the different impressions of heat and cold. It is for the same reason that when one takes a cold bath, the water seems colder that the air, and the air colder than our clothes, although all these objects are of the same temperature. In the same way, again, a child who comes out of a warm bath, and is wrapped up in towels of the same temperature, feels all the same a chill; it is because air and stuffs transmit heat more slowly than water.

## VII.

### ERRORS OWING TO MORAL CAUSES.

Hitherto we have only considered illusions and errors arising from physiological or purely intellectual causes. But for children, as well as for adults, there are also moral causes of error. So many sentiments and inclinations, so many sources of error. It would take too long to pass them all in review, so we will content ourselves with a few prominent examples.

One frequent cause of error in children, is their feverish need of movement and action, their passionate instinct of exercise and of play. They often observe very well (but only up to a certain point) things which interest them, because they pay attention to them. But it is not a search for truth which interests them, it is the need of acting, and of acting quickly to arrive at such-and-such an end. Thus they form their decisions, so to speak, by inspiration. They do not trouble themselves about the real bearing, or the limits or the distant results of things ; they have seen, or half-seen what they wanted to see—the means of producing a certain result, and that is enough for them ; they believe and they act. Too much precipitation is as detrimental to action as to thought, as they often find out at their own cost, but only to fall back again soon into the same errors. Experience comes slowly. Here is an example of this kind of errors, which at the same time marks " the intimate connection of thought and of language, and the mutual help they give each." With children, as with animals, actions which appear to us sensible, make us suppose them to be more intellectually developed than they are. When a child, for instance, perceives that the pin which fastens the wheel of his little carriage is loose, he soon comes to foresee that the slightest shake will cause it to fall out, and that then the wheel will fall also. From that moment the idea of making the pin secure naturally presents itself to him ; but we are not sure that there is anything in his mind but a simple succession of images. By his action he realizes the image of a linch-pin and a wheel securely fastened, because he does not wish for the continuance of the other image in which these same objects in their unsteady condition presage a catastrophe. In order for him to set to work to seek out the causes, which would be a real exercise of reason, it would be necessary for his intellect to make another step in advance.

" Without presuming to decide that this progressive step depends entirely on language, I maintain at any rate that words will be the means of establishing it. Ask a child to give you his reason for what he did to mend his cart. The words which he will use in reply, will lead him to think that he acted thus in consequence of certain laws which may be applied to many things besides his cart. The particular image will become effaced in the presence of more general ideas. He will tell us that, the pin being loose, the wheel would certainly fall ; and this word *loose*, the designation of a quality separable from the object itself, applicable to a thousand other objects, and suggesting a necessary event, has already become a germ of general ideas. Notions of the effects of weight, of friction, in short, of all the immutable laws of nature, will in time result from this imperfect explanation which he has given us."

Vanity, self-consciousness, and obstinacy also lead children into a number of errors. Marie is on the swing, where she makes more noise than all the three children who are looking at her and awaiting their turn. In five minutes, she manages to get in more than twenty phrases, all beginning or ending with the word *I* or *me*. "Look at me, how fast I swing myself! No one can go as high as I can! Look, I am swinging myself all alone!" She fancies all this is true, whereas none of it is : it is illusion of judgment produced by an exaggerated feeling of personal skill. "My pleasures, my troubles, my pains, and my happiness," are all exaggerated by the child, through inexperience, through want of foresight, and through a strong sense of personality.

It is needless to enlarge on the perpetual illusions and errors which self-interest, especially when connected with the palate, give rise to in children. They are patent to every one. But it is curious to observe the facility with which their tendencies or their self-seeking habits lead them unconsciously, from the best intentions, into error, and even into the practice of sophistry. Two children, one seven and the other four years old, were often very quarrelsome together, either from jealousy or thoughtlessness. The younger, formerly rather ruled over by his brother, is now no longer afraid of him, and often provokes him. It is only by fits and starts that they are kind to each other, and then they say, "Shall we be friends?" The treaty concluded, it lasts as long as one could expect. The other day we were witness of a very droll scene. Some one had promised at dessert to divide a piece of sugar between them. The younger one said to his mother, "Mamma, I want to be kind to Charles; you must give him the biggest piece." The mother, however, cut the sugar as equally as possible, and there was hardly any difference to be seen between the two pieces. There was a slight difference however, which both children noticed. Charles said, "They are the same size," preparing at the same time to take the larger bit. But Fernand was quicker, and presented him with the smaller piece, saying to him: "Here you are, Charles, here's the largest piece." And Charles, not daring to contradict him, but very much vexed, accepted it without a word.

# CHAPTER XI.

## ON EXPRESSION AND LANGUAGE.

### I.

HUMAN language, or speech, is only a superior kind of the faculty of expression which all, or nearly all, animals possess. Birds by their attitudes and their song, dogs by their barking and howling, ants by the contact of their moving antennæ, certain kinds of fish by sounds, probably related to their instinct of propagation, in short, the greater number of animate beings are able to communicate to each other, species with species, and individual with individual, by ocular, auricular, or tactile signs, their sorrows and their joys, their desires, their love, and their anger. For man and for other animals, language has the same origin—a complex origin, and one in which direct observation of little children at the period of lingual evolution may give us much valuable light.

The principle of language, or of expression in general, is the correspondence of certain organic movements, strongly marked outwardly, with the inward sensations and sentiments experienced. The internal modifications of the machine are revealed by peripheric modifications, as constant as they are varied in each species. Such are most of the movements, the sobs, the laughter, the babbling of children, mechanically executed from the first months. They necessarily become conscious of them after they have performed them a certain number of times ; but they attach no significance to them, and produce them at first without the slightest intention. These are neither more nor less than reflex actions of the organism. But, with intelligent beings, these unconscious utterances soon develop into

signs. This result is the work of the association of ideas.
The animal or the infant learns to associate the idea of these
organic phenomena,—guttural or other sounds, movements of
the limbs, screams, sobs, laughter, and tears,—with the idea
of the accompanying sentiments or sensations. By a kind of
natural selection these movements are changed from spontaneous to conscious, from mechanical to voluntary. The
child, who produced them at first from sheer necessity of
his nature, now repeats and perfects them for pleasure or
use. An infant of three months, who makes intentional
gestures with his little arms in order to ask for or to repulse
some object, knowing by experience that these gestures will
be understood, already exercises the innate, organic, and
hereditary faculty of expression. When he cries in order
to get his feeding-bottle, to be taken up, to be held in the
arms, to refuse a disagreeable dose, the action, which was at
first automatic, has now become conscious and intentional;
it is even so much a matter of habit that it might be called
reflex. When tears of this sort are too promptly attended
to, and the desires expressed by gestures, screams, or other
signs too readily granted, children are apt to repeat the
process from caprice and habit, without desire or intention;
they do not always shed tears without a motive, but it
happens very often.

The best proof of how much natural affinities and hereditary influences have to do with the first progress in speech,
is, that little children appear, from the very first, to understand the simple language of their mother, and can distinguish their different tones of joy and anger, of coaxing or
threatening. Tiedemann says of his son at the age of one
month: "When any one spoke to him he tried to produce
sounds, very simple ones it is true, and quite inarticulate,
but nevertheless varied." A little girl, three weeks old, who
is by my side as I write these lines, stops screaming and
crying instantly when her mother speaks coaxingly to her.
I have also often noticed that quite young kittens will begin
to whine if left for a moment alone; but they stop directly
they hear their mother coming back. But to return to our
children. A little girl who began to smile at fifteen days old,

used also at this early age to express by particular sounds that she wanted to be fed. At three months she would utter little cries of joy and admiration when she saw flowers or birds, or any brilliant or moving object. If her mother called her attention to things of this sort, saying : " Look at this pretty flower ! " " See pretty Coco," or " this pretty thing," she would instantly begin this joyous babbling, accompanied by little graceful actions, expressive of desire, admiration, and delight. She attached a meaning to a number of words repeated to her by her mother, turning her head towards the dog, when told to " Look at Médor ! " or toward the bird-cage at the words, " See Coco." No doubt also she attached some vague sense to the varied sounds by which she expressed joy or anger, or violent desire. But she did not appear to produce these with the same decided intention that her movements or her screams, sobs, and tears indicated at certain moments.

A child of seven months, who had never before seen me, smiled at me as if I were an old acquaintance, on hearing me pronounce his name. Here at any rate we have an instance of a vocal sign joined to the memory of caresses lavished on him by his nurse or mother while calling him by his name ; or else perhaps the sign is associated with the idea, already confusedly conceived, of his own personality. When I called him by his name, or gave him something he wanted, or if he procured for himself some much-desired object, he would instantly turn to his mother smiling ; or if he was frightened, astonished, or uneasy, he also turned to her in a significant manner. It was as if he wished his mother to know when he was happy or in trouble. At nine months he uttered little cries of pleasure, or cries to attract attention, some of which were evidently imitated, at the sight of a dog, a cat, or a bird, and even once of a fly which had crawled on the edge of his plate, and which his eyes followed for a long time after it had flown away. At eleven months he understood the meaning of a great many words, and even of a few little phrases ; and he used to gesticulate with much expression. One of his intentional gestures, which consisted in clasping his two hands together, interested me very much. He would stand upright in front of an arm-chair against which he supported himself slightly; his mother, three steps behind him, would show him a piece of cake, or something that he had wished for. If she did not bring it to him he moved his whole body round, making eloquent signs that he wanted some one to help him to walk to her. His mother would wait a little longer, and then he would join his hands several times successively, crying at the same time. He had been for some time in the habit of making this gesture when he very much wanted anything which he either noticed of himself, or which was shown to him. His parents liked to designate it as the gesture of supplication, which it would be very easy to explain in his case as the result of heredity, and which it is impossible to refer to the influences of imitation, the action never having been per-formed before him. It has seemed to me that the action of mechanically bringing the hands together to seize or to retain an object, may have remained even when the object was refused; in fact, in the violence of desire, imagination tends to confound the idea and the sight and the possession of a thing.

This gesture, at first unconscious, may easily become intentional on the part of young children, and develop into the gesture of prayer, provided that their desires are satisfied when they make it. The child I have been speak-ing of had nearly arrived at this point. I think this remarkable gesture should be connected with analogous movements that are so frequently seen in young animals. A kitten, six weeks old, seeing its mother and two bigger sisters standing on their hind paws and catching with their front paws the food that was being thrown to them, tried to imitate them; but she tumbled awkwardly on her side without accomplishing any result. Then she took to climbing up our trousers or petticoats to get at our hands. A few days later, having become a little firmer on her hind quarters, she made another attempt at catching her food, and half raised herself, bringing her front paws together as she saw the other cats do. Often, however, the movement of prehension was made before the bit of food had been thrown or after it had fallen, or been caught by more skilful hands, the kitten's paws still remained clasped together, as if the wished-for prize had been obtained. As in the case of the child above mentioned, the ideas of the desired object and of its possession get confused : the gestures are finished in the air.

Let us take another child of eleven months. He has a little cardboard horse on wheels, which he treats as a sort of scapegoat. He knocks it about, strangles it, pinches it with all his might, etc., etc. These interesting exercises are habitually accompanied by the exclamations, *Hue, hue,* which he learned about a month ago. When any one says to him, " Beat the horse," he takes his little wooden spade, and belabours the back of the insensible quadruped. A lady friend of his often sets the child astride on her knees, saying ; *broum / broum /* words which the little rider imitates as well as he can, while performing his imaginary gallop. He also says distinctly, *papa, papa,* without attaching any sense to the words, as he has not seen his father, even in a photograph. He has been taught to make with his hand the gesture which corresponds to the words *au revoir.* He cannot yet pronounce these words, but he makes the gesture mechanically, through simple association of ideas, and only when he is told to do so, to any one who is going away. He also performs a little comedy of his own. If one says to him, *Prise,* he sniffs like a person about to sneeze, with an irresistibly droll curl of the lips and the tip of his nose. This is the first act, which is followed by a second more complicated and not less curious. An empty snuff-box is given to him into which he inserts his fingers, carries them to his nose, and recommences the performance of which we have just spoken. The more one laughs at him, the more ceremony he makes about conveying his fingers from the snuff-box to his nose, the more he sniffs and turns up the tip of his nose. He understands the meaning of such phrases as : " Give this ;" " Take that ;" " Drink this ;" " Eat that," etc. He has thus acquired not only knowledge of signs representing individual objects, but, as we can see, he also understands signs representing fairly complex ideas and actions, and a few of which he has succeeded in imitating. He understands the value of a word, and every day he makes fresh progress in the art of speaking.

## II.

Thus then we see imitative and oral language developing side by side; the former, however, thrown back for a time because of the rapid progress of the latter, which is the necessary, universal, and, so to say, official instrument of human expression. This is also the more interesting language for the psychologist to study, and it will be convenient here to give it special attention. Let us go over briefly the first lines of this branch of study, and, checking our own observations by those of eminent observers, let us attempt in broad outlines a sketch of infant glossology.

First of all, it is necessary to decide which are the different parts played by instinct, organization, heredity, and education—*i.e.* imitation—in the acquisition of language. Man, like other animals, utters sounds and produces spontaneous movements which soon become conscious and voluntary signs of what he feels and thinks: from the moment of his birth he begins to speak the language of nature, and possibly the language of his forefathers. "The intimate connection between the brain, as it is now developed in us, and the faculty of speech, is well shown by those curious cases of brain disease in which speech is specially affected, as when the power to remember substantives is lost, whilst other words can be correctly used. There is no more improbability in the effects of the continued use of the vocal and mental organs being inherited, than in the case of hand-writing, which depends partly on the structure of the hand, and partly on the disposition of the mind ; and hand-writing is certainly inherited." [1]

The part played by heredity in language, side by side with that of imitation, and which appears to us so consider-

---

[1] Darwin, *Descent of Man*, vol. i., p. 58.

able, is far from being established by experimental proofs, either in the case of men or of animals.

"The experiment has never been fairly tried, of turning out a pair of birds into an enclosure covered with netting, and watching the result of their untaught attempts at nest-making. With regard to the song of birds, however, which is thought to be equally instinctive, the experiment has been tried; and it is found that young birds never have the song peculiar to their species if they have not heard it, whereas they acquire very easily the song of almost any other bird with which they are associated."[1]

"The sounds uttered by birds offer in several respects the nearest analogy to language, for all the members of the same species utter the same instinctive cries, expressive of their emotions; and all the kinds that have the power of singing exert this power instinctively; but the actual song, and even the call-notes, are learnt from their parents or foster-parents. These sounds, as Daines Barrington has proved, 'are no more innate than language is in man.' The first attempts to sing 'may be compared to the imperfect endeavour in a child to babble.' The young males continue practising, or, as the bird-catchers say, recording, for ten or eleven months. Their first essays show hardly a rudiment of the future song; but as they grow older we can perceive what they are aiming at, and at last they are said 'to sing their song round.' Nestlings which have learnt the song of a distinct species, as with the canary birds educated in the Tyrol, teach and transmit their new song to their offspring. The slight natural differences of song in the same species inhabiting different districts may be appositely compared, as Barrington remarks, 'to provincial dialects;' and the songs of all allied though distinct species may be compared with the languages of distinct races of man."[2]

We see in like manner that human beings only speak as they have been taught; but it would be interesting to know

---

Natural Selection, Essays by A. Russel Wallace, p. 220.
[2] Darwin, Descent of Man, vol. i., p. 55.

if, in the existing stage of the development of the human brain, they would not of themselves evolve a sort of hereditary language, of which the first efforts of infant language would be the basis. The experiment made, according to Herodotus, by king Psammetichus, is not, in this respect, either sufficient or even authentic. The king of Egypt, curious to know what was the most ancient nation of the world, had two new-born children shut up in a cottage, where they could hear no one speak, and where they saw no other living beings than the goats who nourished them. At the age of two years, when some one went into their room, these two little savages uttered the word *beccos*, which signifies *bread* in the Phrygian tongue. Psammetichus, so runs the legend, then thought himself sufficiently authorized in proclaiming the Phrygians to be the most ancient people of the world. We shall be more exacting than this royal experimentalist. We declare ourselves to be absolutely incapable of deciding *à priori*, whether children subjected for a sufficient length of time to a like treatment would work out for themselves anything like a real language, what sort of language it would be, and what sort of ideas would be evolved in their brains, abandoned solely to the resources of hereditary transmission and imitation of nature.

Most of the recorded facts relating to the sequestration of human beings at a tender age, and which some philosophers think conclusive, we regard simply as not having happened, as far as science is concerned. ·

"We read sometimes in the newspaper, of the brutal state of those unfortunate beings whom avarice or cruelty had from their early childhood kept confined in dark cellars, and excluded from all social intercourse and intellectual excitement. The physical and mental state of such individuals is mere vegetation, not a developed human existence. . . . It was impossible to impart to the well-known Caspar Hauser the idea of a horse. When the word was pronounced, he thought of a wooden toy which he played with during his imprisonment, being unable to attach any meaning to the word horse but in this connection.

. . . Corresponding observations have been made on men, who from their earliest childhood had lived and grown up among animals, removed from the society of men. They lived and supported themselves like animals, they had no spiritual wants ; they could not speak, and exhibited not a trace of that 'divine spark,' which is said to be innate in man."[1] The observations of this well-known German materialist have no semblance of scientific experiment. That an infant hidden away in a dark cell, that a young savage nurtured by goats, and who actually lived and grew up in the woods, should not be able to speak, or show by the ordinary means of those who can speak, that they have any general or specific ideas of existence, is not at all surprising. But is it certain that they did not possess ideas —even general ones—though they had no means of expressing them ? If five or six children were shut up together, would they not evolve a common language ? This has not yet been tried. This experiment, easier to make and less barbarous than might be supposed, would no doubt enable us to arrive at something more than conjectures on the origin and nature of language, and on those interesting problems of psychogenesis, which contemporary philosophy approaches with so much reserve and so few results, but of which natural science is preparing the probable solutions.

The ideas propounded by M. Taine on the acquisition of language by children are a mixture, somewhat confused at times, of the theory of Condillac and the modern theory of selection.

According to him, children begin to exercise their vocal organs by a continuous series of cries and varied exclamations ; for several months they do not get beyond vowels ; "by perpetual groping and attempts, and by gradual selection, consonants are added to the vowels, and the exclamations become more and more articulate." At twelve months old, his daughter appears to him to have acquired the material of language, without attaching any sense to the sounds

---

[1] Büchner, *Force and Matter.* See English translation by J. F. Collingwood, p. 164.

which she **uttered.** " All the initiative," he says, "is her own ; it is personal, accidental, and momentary invention which have caused her to find and to repeat such sounds as *mm, kraaau, papapa ;* example **and** education have only served to direct her attention **to** sounds which she had either attempted or found out for **herself."** I must say that in my opinion the list of **sounds originated by** children is a very limited **one, and** little **adapted for the** basis of any **solid argument, touching the** respective *rôles* of personal **initiative and of imitation in** the acquisition of language.

Pleading **still the** cause of the spontaneity **of** invention or re-invention, which precedes in children the **work of as-** similation, M. Taine affirms (no doubt with **reason) that** this infantine babble is of astounding flexibility and variety of meaning, and that every shade of emotion,—astonishment, gaiety, vexation, sadness, **etc.,—is expressed in its** various tones.

Mr. Darwin also contributes some observations on **the** means of communication possessed **by** children. These are, at first, instinctive cries intended to express suffering in general ; **after a** time, **the tones vary** according as pain or hunger is the cause of the utterances ; **a little** later, the power of voluntary **tears is** added ; at the age of forty-six days, there are little varied sounds produced, as if by pleasure ; at 113 days, we see the first attempt at a smile, and possibly at the same date efforts to imitate sounds. The articulate sound *da* is formed at **five-and-a-half months.** " When a year old, he tried to invent a word to designate his food, and produced the sound *mum ;* and henceforth this syllable sig- nified : " Give me something to eat." **This word** corres- ponds to the *ham* used by M. Taine's child, so **Mr. Darwin** tells us, and he seems to attribute it wholly to spontaneity and not to imitation, for he adds that he does not know " what led the child to adopt that particular syllable."

M. Egger notices at five **weeks old the** transition from the cry to the voice. The cry is the first sound that the human organ **makes ;** it proceeds from the bottom of the larynx and dates from the moment of the first breath. For **several weeks it is the only sound** a child can make when he is in pain. "Towards the fifth week I observe the mouth and tongue beginning to move (especially when joy is felt), and to produce sounds which cannot be represented by any letters in our alphabet, but which are certainly less guttural than the previous sounds. These last sounds, as they go on perfecting themselves, become real articulations. . . . With Felix this distinct utterance of a cry of pain was scarcely heard before the end of the second month."

As regards the infant voice, M. Egger makes one observation which does not appear to me to be borne out by facts ; it is, "that the voices of children in earliest infancy are not characterized by an individual *timbre*. The voice acquires a distinct character at the same time that it becomes articulate, and that we can distinguish vowels and consonants in it." I have no contrary observation to oppose to this statement, but I read in a pamphlet of Dr. Laurent's, that "the particular *timbre* of the infant's cry varies like the human voice. In each child it exhibits particular modifications, which language cannot express, but which the ear can seize ; and a mother can distinguish her own child's cry from that of others." This point then, more interesting to the philologist and the naturalist than to the psychologist, calls for fresh information.

M. Egger, like M. Taine, assigns a large share in the development of language to personal initiative and invention. Besides the natural language of tears, cries, smiles, and gestures, "which becomes the commencement of artificial language when *intention* is mixed with it," he admits the involuntary play of the voice, which, from the age of six months, causes infinite variety in the attempted sounds and articulations. Thus we seem to have an instinctive, natural language, common to all times and all people, which recedes gradually before the progress of another language, invented by the child and susceptible of a host of individual variations. Neither M. Egger nor M. Taine have noted the forms of this language, which, according to them, is spontaneous. Their observations have too vague and general a character for their conclusions to be definite. One must undoubtedly admit, even in the first period of infant

babbling, that there is an hereditary instinct which urges
the young nurseling, like the young bird, to try its voice,
according to the measure of the feeble resources of its
organs, for pleasure rather than necessity, and from chance
rather than voluntary imitation.   But already, even at this
period, involuntary imitation, organic sympathy, and musical
contagion have their part in the work.   As to this first
period of artificial language, particular to each child, useful
for inter-communication with other children, and especially
with his nurse and his parents, it seems to me rash to affirm,
without the slightest experimental proof, that " there is not
one of his wants for which he does not invent one or several
articulate sounds, without any example being set him, intentionally or otherwise."

One of the most competent observers, M. A. de la Calle,
has just made a notable contribution to the store of published
observations on the formation of infant speech.   According to him it is very difficult during the first three months
to observe any appreciable phonetic phenomena.   This
difficulty, however, has not discouraged him, and we
have his permission to quote some of his important observations.   In one of his children he noticed on the forty-fifth
day a little bleating sound which accompanied his smiles ;
and from this date the child seemed to take pleasure in
exercising his phonetic organ.   " At this period of life," he
says, "the faculty of imitation has not yet shown itself, and
one cannot admit that these babblings represent attempts to
imitate any sounds heard, or the songs with which the nurses
lull them to sleep."   He attributes the pleasure which the
child finds in this babbling, which is as yet neither expressive nor musical, rather to a combination of reflex actions
associated for the first time with conscious effort.   He has
moreover remarked that his children produced these babblings with varied tones, more particularly under the influence of some agreeable impression ; he adds with reason :
" What is certain is, that these sounds, reiterated and more
or less continuous, have more affinity with the sounds of
laughter than of tears. . . .   Natural selection ought to
bear principally on the sounds expressing agreeable sensa-

tions, in virtue of the principle that "all pleasurable states mean an augmentation of part or the whole of the vital functions."

M. de la Calle has endeavoured to arrive at an exact summary of the variety, number, and value of the different sounds which a child utters before the appearance of anything approaching to an articulate syllable ; and though he could only succeed, he says, in obtaining a statement of the fact, he endeavours to determine the eight different sounds which compose the musical scale which a child goes through several times in the day. He notes in the first place the apparition of the vowel *a*, which he calls the vocal or sounding cellule ; the sound *ai*, *é*, is produced next, and sometimes at the same time, as *au*, and *ô*, which are only the first differentiations of the primitive sound *a*.

"At the age of six months, Fernando used to execute a very characteristic series of sounds ; with the help of this single sound *a* he could sometimes keep up a conversation for a quarter of an hour with his mother or nurse, laughing, gesticulating, and babbling all on this same vowel, which he varied with infinite shades of meaning, with modulations and particular inflections, and with a fluidity, so to say, which our organs, stiffened with use, would be altogether incapable of. The intensity, swiftness, duration, and pitch of the sound were varied by insensible gradations, with alternations and repetitions excessively curious to study.

"A little later, at eight or nine months (and now he had the additional help of physiognomic and imitative expression), he still used by preference this same vowel *a*,—though he already knew how to articulate several syllables,—for pointing out, naming, calling, and asking for all sorts of things and objects : *àh* (inspiration) signified astonishment and content ; *àh* (expiration), a call or indication of all things and objects ; *hà*, *hà*, *hà*, very short and repeated, to ask for something which he saw and wanted ; *hâ*, *hâ*, sustained, when he thought something very pretty, and to express admiration generally. *Hâ*, *hâ*, *hâ*, to intimate a wish that some one should hold his hand to walk ;

sometimes too that he wanted to be taken out of doors. Gestures, and the expression of the countenance came in noticeably to the help of his interjections, the child pointing out the desired object, or else extending his hand and contracting his fingers several times as if to seize it ; the direction of the eyes and the expression of the look, the movements of the body,—in short, all the expressive manifestations that we have already studied,—are used by children with more or less skill.

I have also remarked that in certain cases the sound *ai* was produced at the same time, especially on the higher notes ; but I have not been able to distinguish *o* very pure in any of my children. It was only when they were crying that I ever noticed it, and then it was like a compound sound or diphthong ; *o* absolutely pure I have only been able to observe in the articulate voice, before or after eight or nine months, according to the individual. The following have seemed to me accurate transcriptions of my children's crying sounds. After the third month [the sounds are French] *à-há, à-há, à-há, è-hê, hèu !! áh, áh, œu !! oheu, à-há, à-há, à-há, oheu ! è-hê, è-hê, è-hê, à-há*, etc., etc. But these cries are not those of the sharp grief, which is expressed by *àh* or *eh !* . . . always deep and screeching.

"This preference, eminently physiological in children, for the vowel *a*, inspired or expired, as the case may be, is significant of the part that it plays in all rudimentary and primitive tongues. It is the simple cellule, the phonological embryo of spoken language."

It would take too long to give even a simple *résumé* of all the observations made by M. de la Calle on the first phonations, on the vowels, diphthongs, gutturals, and labials. We must be content to mention the principal phases of this evolution.

"At four months and a half, he articulated the first explosive sound joined to the phonetic *a*, saying *àp' pá, àp' pá*, with an effort, which was evidently very great, for he opened his eyelids wide and his eyes became very bright.

"Some days after, he produced the sounds ' *mam' ma, mam má* with less effort, or I should perhaps say with an

250 THE FIRST THREE YEARS OF CHILDHOOD.

effort already better localized in the organs of speech; for
though his eyes again sparkled, his eye-lids did not move;
the effort is here characterized by the nasal sound itself, as
if the child wanted to speak and yet could not. It is in
fact the *mu-mu* of the deaf and dumb, by which they inform
us that they are not able, or do not know how to speak.
The English have a word *mum*, which means " Hold your
tongue !"

"At five months he had in a slight degree conquered
the difficulty, and could accomplish the articulation during
the expiration of a breath, and he could say *pàpá* (sharp),
*màmá* (a little less sharp), but both words were very short
and staccato.

When he was in a passion, he gave out a prolonged
staphylin-guttural cry, clenching his fists and getting very
red in the face; one would have said it was the cry of a
monkey; but when he felt very joyous, or if any one
tickled him for any length of time, and his laugh became
spasmodic, the sound was prolonged from the expiration to
the inspiration of the breath, and he then produced a sharp
*khr-ahr*, exactly like the sound of an animal. I myself
provoked the repetition of this performance two or three
times, in order to be quite certain about it; then he put
himself in a rage again, pouted his lips, began to cry, and
I was obliged quickly to kiss him in order to gain pardon
for my strange behaviour. Often, when he was happy and
satisfied, but not over-excited, he would articulate the
vowel *e* with a little explosive sound, medio-lingual, medio-
palatal—*ké-cké*, and he would amuse himself by repeating it
for some time while he was apparently carefully examining
his fingers and hands which he turned round and round, or
else inspecting all over some object which had been given
him to play with.

"The modulations were often changed unconsciously and
without signification; it was a vocal exercise which he
practised for pleasure, and which had gradually taken the
place of the babblings of earlier days; but considerable
progress had been made by means of these articulations,
for the child now possessed all the materials for words.

" The sound *ké-cké* became now *ká-cká*, now *ak-ká*, **now,**
*tà-tà-tà, à-tá-tá, à-bà-bà-bá, dá-dá-dá,* and **then** *pé-pé-pé,*
*bé-bé-bé,* etc.  A kind of gentle plaint, either when he was
worried by being dressed, or in any other way,—for instance,
if he was hungry and was made to wait a little for his food ;
*ayá' ya ! aya' ya !* **aya' ya !** was a differentiation which I
think I have only remarked at nine months old ; nevertheless I seem to have heard it a good deal sooner in my
other children.   However that may be, this discovery was
very important to the child, and he was probably aware of
it, for he enriched his language with a greater variety of
compound sounds while making use of this new scale.

" Thus, *há-há, hé-hé, ká-ká,* which were already a great
advance in facility of articulation during expiration, were
varied by differentiating and lending themselves to new
phonetic combinations; *há há* became *oyá' yá ; hé hé,*
*eyé eyé ; ká ká ; kayà' yá kayá' yá,* and so forth ; and the
song would begin again with these new variations."

### III.

We have now come to the period of the most important
acquisitions of words.

For more than a fortnight I have been observing a child
of twelve months who is backward in walking, and can only
toddle along when held by the hand, but who is very precocious in talking.   A month ago he could walk alone, but
he had several falls and he is now afraid to repeat the
experiment, and if any one tries to make him do so he sits
down on the ground and says ; *gnô, gnô (no, no).*   He
knows the meaning of a large number of words which he
hears, and he uses some of them himself in the ordinary
sense ; he says *painm* to ask for bread (*pain*) ; when his
frock is being put on, and he sees his hand coming out
through the sleeve, he calls out *mainm.*   His brother Charles,
whose games and mischievous tricks delight him, he calls
at every instant : " Kiah, kiah."   Meat (*viande*) he calls
*miamiam ; mené mené mené,* indicates the desire to go from
one place to another ; *peudu (perdu)* indicates an object

that has fallen, or been thrown away or disappeared ; sometimes he says *a pu* (*il n'y a plus*) to express that a thing
is either eaten, finished, hidden, or taken away.   He imitates
donkeys, cats, cocks, and especially pigs ; when he wishes
very much to be taken out by a particular person, and has
gained his desire, he expresses his great pleasure by mimic-
ing that interesting animal.   The other day, I was very
much  astonished  at  hearing  him  reproduce  the  first
syllable  of  a  word  spoken  in  his  presence.   "Will  you
have some cake (*gâteau*)?" his nurse had said to him, and
he instantly answered : *Ga, ga*, and directed the attention
of his nurse towards the dining-room, the use of which he
already knew, although he had only been in the house two
days.   He also knows the meaning of many of the simple
phrases which are used in speaking to him.    There is no
occasion to give examples of these, which are analogous to
those already quoted.    But what is more important, I heard
him the other day attempting a thoroughly synthetic phrase
by putting two names together; his accompanying tones,
gestures, and movements left no doubt of his meaning.   He
had been walking with me for several minutes, when he
saw his mother and  wanted to go to her, and as I did not
instantly take him to her, he pulled hold of me, pointed to
his mother, and with a supplicating look said several times :
*papa—maman* (papa, lead me to mamma).  A little later, and
he would have interposed the word *mené* (*lead*) between the
two names.   The advance however was not made this same
day, perhaps because I had not the patience to wait, but
obeyed the child's wish too quickly.    Besides the movement of the head from left to right to signify *no*, and that
of pointing out an object with the hand, I have noticed
another very characteristic one.   When any one sings or
plays the piano, he sways backwards and forwards rhythmically and almost in time ; he has been accustomed to
do this since he was five months old ; and he sometimes
accompanies the movements with a musical *brumm*.

In another child, thirteen months old, I have noticed an
expressive gesture which I have not yet quite made up my
mind about.   His parents assure me that they have not

taught him a movement of indication with the forefinger, which might be hereditary, or might proceed by selection and simplification from the act of prehension. Here is another action whose evolutionary origin does not appear to me to be absolutely demonstrated. In the same child, the gesture of negation, which consisted in waving the hand backwards and forwards, may be considered in principle as a derivation of the action of pushing away any disagreeable thing. Two months ago there was no difference between the movements with which this child pushed away anything on the table which he did not want, and his refusal of anything that was offered to him which did not suit him. At present he uses this gesture in the following way also ; if he sees some one out of doors who reminds him of his father or mother, he considers the person attentively, and after perceiving his mistake he says : *mamma, papa,* accompanying the word with this gesture of refusal or repulsion, which is his way of saying *no.* The gesture of the head meaning *yes,* he made at a very early age, I am told, without having seen it made. It is a gesture evidently reflex, like the half-shutting of the eye-lids which accompanies it.

A little girl, nineteen months old, has already accomplished three-quarters of this important and relatively rapid process of evolution, which is the initiation into language. She cannot yet say any phrases, however short, though she understands the meaning of a great many tolerably long ones ; but she pronounces a quantity of words intelligibly. There was no difficulty in making her pass from inarticulate to articulate sounds ; she felt after them instinctively, but she needed the help of imitation to enunciate them easily and distinctly.

Some months ago, when she was beginning the acquisition of her vocabulary (now a very rich one), she could only reproduce the last emphasized syllable of a word ; and she used to alter its articulation according to the rule of what was least effort. It is noteworthy, that in the language of children, as of primitive races, the roots or first attempts are monosyllabic sounds ; in young animals the voice itself is only a single guttural or labial intonation, prolonged or repeated at short intervals. The puppy does not at once give forth the characteristic bark of the dog, which the French represent as *wrroua* (with a guttural *r*), nor the kitten the *miaou* of the cat, nor the young house-sparrow or swallow, their *kiou-kiou* or *tyi tyiri*. Their first vocal attempts are rudely simple and monosyllabic. The same with children; they more easily pronounce reduplicated monosyllables, as in *papa*, *mama*, than dissyllables such as *gâteau*, *minet*. They will sometimes even bring out diphthongs, which are varieties of monosyllables like *oua*, *mia*, *mié*, etc. For a long time they rebel against real dissyllables, and still longer against polysyllables. For a long time this little girl, whom I have observed ever since her birth, up to the twenty-second month, but unfortunately without noting each step of her progress, could only say *bou* for *tambour*, *fé* for *café*, *yé* for *Pierre*, etc. By perseveringly separating each syllable for her very clearly, her mother at last succeeded in making her pronounce the distinct syllables of the words. The mother repeated slowly and distinctly, *tambour*, *Georget*, *pomme*, *pain*, *gâteau;* and after a considerable number of fruitless lessons, the little pupil succeeded in articulating, *a-bou*, *o-yé*, *om*, *pai*, *a-teau*, etc. It was the same with a number of other words, which she now pronounces distinctly enough, save for the modifications of the consonants, which vary with every child. There are no rules in children's pronunciation of the sounds which they imitate. A child of fifteen months who says *a-to* for *bateau*, for *gâteau* says *ca-co* or *cacou*, or even *a-ca-cou*, when in a very voluble mood. He says *tu-tu* for *tortue* (tortoise), and also for *confiture* (jam). The same syllable pronounced in one way, in one word, is often pronounced differently in another word, or doubled, or augmented by a useless syllable.

In general, we see clearly in infant language, the application of certain rules which philosophy has long recognised in the transformations of language ; and among others, the modifications produced by the tendency of individuals to diminish the muscular efforts of pronunciation. It is a very common thing in France among French people to hear *particulier* pronounced like *particuyer*, *cuiller* like *cuyé;* and this tendency to pronounce *ll* like *y* is on the increase, by virtue of the law of the *least* effort.

In like manner children begin by pronouncing the consonants which are easiest to articulate, and they modify them little by little in proportion as their organs become stronger and more practised.

Mr. F. Pollock has been one of the first to try and systematize in phonetic as well as logical order the first infant gropings after language. His interesting study on the " Child's Progress in Language," comprehends the period between the twelfth and the twenty-fourth month. We quote here the most essential points :—

" Age, twelve months. *M-m* often repeated ! *bá bá* repeated an indefinite number of times.

" *M-m* generally indicated a want of something. *Bá bá* was (1) a sort of general demonstrative, standing for the child herself, other people, or the cat (I do not think she applied it to inanimate objects) ; (2) an interjection, expressing satisfaction. Both sounds, however, seemed often to be made without distinct intention, as mere exercise of the vocal organs.

" Thirteen months. *Dá dá* was used, at first as a vague demonstrative (and about six weeks later it became a distinct proper name for the child's father). *Wa wa*, meant water, drink ; *wah, wah*, with a guttural sound distinct from the foregoing, was said to figures of animals—dogs or cats, which she now recognised in pictures. This fact is curious, having regard to the inability of adult savages, as reported by travellers, to make anything of even the simplest representations of objects. *Nà-ná*, signified nurse—of course as proper, not generic name.

" Fifteen months. *M-m* discontinued. *Bá bá* was sometimes used instead, and sometimes she simply cried for a desired object. *Wah wah, miau,* soon became generic names of dog and cat (*wah wah*, which at first included cat, becoming appropriated to dog). I think, however, *wah wah* would include any middling-sized quadruped other than a cat or a sheep. As to cat, her name for it became a few months later *aya-m* or *ayá-m*, which, so far as I know, she invented for herself. The conventional 'gee gee' for horse was very soon understood by her, though she could not form the *j* sound. She recognised a zebra in a picture alphabet as 'gee gee' and showed marked dissent when told it was a zebra.

"These imitative sounds were all learnt on the suggestion of adults, but studied from the real sounds; for, as made by the child, they are decidedly nearer to the real sounds than the *baa baa*, etc., used by adult voices.

"'Baby' (or rather *bê bi*), was now formed with fair success, but soon dropped for a time. About a month afterwards it was resumed, and became the child's name for herself. This was long before she attempted any other dissyllable. It was pronounced, however, rather as a re-duplicated monosyllable.

"Seventeenth to eighteenth month. Her vocabulary is now increasing fast, and almost any word proposed to the child is imitated with some real effort at correctness. The range of articulate sounds is still very limited; *a, á, i* (short and long) are the only vowels fully under command; *â* occurs in a few words, and is the result of attempts to form *o*: thus, *nâ* = nose. The long sound of English *i* (*ai*) cannot be pronounced; when she tries to imitate it, she says *iá* or *í-a*. No approach is yet made to the peculiar short English sound of *a* in such words as hat, bat. Of consonants, *g, l, r* (the true consonant initial sound), the final semivowel, as in *more, poor*, is easy enough to her, and sibilants, aspirates, and palatals are not yet mastered. 'Guy' (a younger cousin's name) is called *dá*, produced far back in the mouth; *k* is also produced far back in the mouth, with an approach to *t*. Final consonants are seldom or never given. With the exception of 'baby' *ná-ni, ná-ná*, the vocabulary is essentially monosyllabic. She once said 'lady' pretty well, but did not take it into use. No construction is yet attempted. But even with these resources the child already contrives to express a good deal, filling up the meaning of her syllables with a great variety of tone, and also with inarticulate interjections. Impatience, satisfaction, amusement, are all very well marked; and perhaps even intellectual dissent (in the case of 'zebra' and 'gee-gee'), see previous page."

Amongst other acquisitions (from fifteenth to nineteenth month) we may notice the word *poor*, with no appreciable difference from ordinary adult pronunciation, and which was taught as an expression of pity, but extended to mean any kind of loss, damage, or imperfection in an object, real or supposed.,

"Nineteen months.—*O* sound now distinctly made, and *g* distinct by the end of the month. 'Guy' is now *gá* instead of *dá*. *L* and *t*, final, and even *p*, are pronounced more or less distinctly. The monosyllabic form still prevails. *K* is a favourite sound, and she has several words formed with it which are kept carefully distinct. *Ku* = stool, *kah* (later *kad*) = cod [liver oil], which she considers a treat. *Ko* = 'cosy' (or tea-pot), *kâ* = cold. *Ká ká* = chocolate, *khien*, or *klien*, = clean. . . . *S, sh, ch, j*, are on the whole indistinct. . . . *W, v, f*, are now formed, but not well distinguished." (Mr. Pollock thinks, with M. Taine, that infant pronunciation has some shades which escape adult ears.)

"Twenty months.—*Dash* or *dásk* = dust. *Ta'sh* or *tá'sh*, learnt, I think, from 'touch' but soon dropped. *Tásh*, however, is adopted for [mous]tache. Final sibilants are more under command than initial. . . . A sudden advance made to dissyllables, several being produced with success on or about the same day. Fanny, honey, money; *fá-wá*, flower; *la-ta*, letters; *ha-pi*, happy; *bá-ta*, butter; *A'-si*, Alice. *R* is still very impracticable, and attempts to form it sometimes give *d;* but this was very transient, and *l* soon became the constant substitute. . . ." (I omit a few other stages of progress of the same kind, to pass on to two or three more important ones, between the twenty-first and twenty-second months). "The child is now more or less able to answer direct, as distinguished from leading questions. Thus, when she had been paying a visit to some relations and cried to go home, she gave afterwards (March 17) a pretty connected account of it in monosyllabic answers. 'What did you do to-day at ——?' 'Klai,' (cry). 'What did you cry for?' '*Ham*,' (home). Also, when told not to handle a forbidden object, such as a knife, she will say, in a tone of intelligent acquiescence; *no—dá-dá* (*i.e.*, I may not have that, but *dá-dá* may). One trisyllable is in common use : *Tenisi* = Tennyson, an illustrated edition which divides her attention with *Vats* (Watts).

"At twenty-one months I noticed a distinct attempt at grammatical construction, by the use of a real predicate so as to form a complete proposition. The child had been told, half in joke, that cabs were dirty as compared with her perambulator. For some days she had been accustomed to say 'dirty' on the mention of perambulator. Now she made the whole statement for herself ; *Kabz dati, klam klín*, (Cabs dirty, peram' clean). . . . (We must also notice progress with regard to generic names.)

"Twenty-two months.—Vocabulary and power of expression are gradually and steadily extending. *Zhátis* is often said for 'There it is ;' '*Out-pull-baby-pecs*' (spectacles); 'Run away, man ;' 'Mamma get Bessie' (her doll). . . . The consonants *ch, j*, and *th* are still imperfect, and consonantal *r* is not yet formed at all . . ." Mr. Pollock also records some expressions indicative of the dramatic faculty in the child, and he adds : "I may observe on this, that I have no reason to doubt that all the play with her doll is purely and consciously dramatic, not animistic ; in other words, I have seen nothing to indicate a belief that the doll is really alive ; nor is there, so far as I can observe, any tendency to attribute life to other inanimate objects. I think the child is perfectly aware of the difference between animals and things, though I am unable to give specific reasons for this impression."[1]

(At this age there appears a decided tendency to imitation of grown-up people's actions, and to asking questions.)

"Twenty-three months.—The palatals, dental aspirates, and the peculiar English short *a* (as in "hat") are still imperfect, and *r* is represented by *l*. When *s* comes before another consonant, one of the two is dropped. *K* is in

---

[1] See *Mind*, July, 1878, vol. iii., pp. 392-395.

some words confused with *p* or *t*. She says *oken* for ' open,'
*kek* for ' take,' . . ." etc.

M. de la Calle also has just added to this branch of study
a useful contribution of personal researches, which are all
the more valuable as he explains the facts collected by the
scientific laws of language. We quote as follows :—

" I have observed, for instance, an absolutely constant
fact : Hard consonants are very difficult for a child to pronounce simultaneously in the same word; he will suppress
the one and change the other, or perhaps replace both, if
they require hard expiration, by an analogous sound with
soft expiration. They will often insert a vowel sound
between the two. This proceeding reminds us of the
theory admitted by philologists, that certain double consonants must have been originally separated by vowels and
afterwards have become reunited, sometimes even fused in
one, like the Greek ψ (psi), and the Latin *x* (gs).

" I have also remarked this same process in the separation of diphthongs by an interpolated consonant, sometimes
amounting to a sort of harmonious reiteration of the altered
syllables ; transposition and contraction are also very frequent. For instance, I find in my journal that Adolphus
at the age of twenty-two months said *cou* for *clou*, *ot-tà* for
*ôte-toi*, *cloute* for *croûte*, *toujer* for *tuer*, *baugrête* for *brouette*,
*liff* for *livre*, *anoir* for *armoire*, *la-lô* for *là-haut*. Then
*rghouise* for *nourrice*, *gouaselle* for *mademoiselle*, *acquelocque*
for *enveloppe*, *cacquette* for *casquette*, *poeterre* for *pomme de*
*terre*. Corresponding phenomena are recorded concerning
the language of the eldest girl. Pepita used to say *lache* for
*vache*, *loture* for *voiture*, *chelal* for *cheval*, *zâme* for *dame*,
*zénêtre* for *fenêtre*, *aristrocate* for *aristocrate*, *pa pi bo* for *pas*
*plus beau*, *les sansan* for *les enfants*. Another little girl
whom I observed at about the same age used to say *les*
*fanfans;* but the harmonic phenomenon in the alteration
was identical.

" The confusion of *l* and *r* is also a characteristic phenomenon of infant language ; children often mechanically
change or replace the one by the other ; but it is the great
difficulty in the pronunciation of the *r* which we are specially

concerned with here. This consonant, pronounced by a vibration of the tip of the tongue against the middle of the palate, requires a certain amount of muscular exercise, and is not mastered till much later. I have noticed that children begin by sounding it in the middle of words, then at the end, and finally when it is initial. But for a long time there will be confusion between these two letters, and sometimes *l* will be substituted for *r*, sometimes *r* for *l*. This reminds me that in cert..n romanized Latin words this tendency to confusion of *r* and *l* has produced some alterations which have been set down to an etymological law, such as *chapitre* from *capitulum ; pupitre* from *pulpitum ; épitre* from *epistola ; apôtre* from *apostolus ;* and, *vice versâ, peregrinus* has become *pélerin*. It appears, also, that the ancient Egyptians did not very clearly distinguish between these two consonants."

M. de la Calle has also gone into the interesting question of the primitiveness of monosyllabism. Contrary to the ideas which obtain at the present day among philologists, he has observed in his three children, " that monosyllabism is no more than polysyllabism, an absolutely constant phenomenon."

The words which children retain most easily are those which express the most salient qualities of things, or those parts of them which produce the dominant impression. Thus, a little girl of twenty months, when I took off my hat and said to her, " What is it?" answered " *bonnet*" (cap). This is her generalized name for every kind of head-gear, masculine or feminine ; it is an object which has often attracted her attention on her mother's head, and on her own, when she looks at herself in the glass. She calls the *carafe, vê (verre),* while a bottle of wine is a *lit (litre)* ; the word *verre* designating a quality which has struck her both in the tumblers and the decanter, while the bottle, although also glass, appears to her under a very different aspect. She also calls a little medicine-bottle which hardly holds a *decilitre, lit.* Her general idea of glass will go on enlarging every day, while that of *litre* will gradually contract as she gains fresh experiences.

A little child of two and a half calls all dogs *oua-oua*, except his grandfather's dog Cambo, although he cannot pronounce this name : he does not class him with the ordinary *oua-ouas*.　He also gives this name to all his wooden toy-animals—the dog, goat, wolf, hyena, and lion.　But when I called a stuffed lion an *oua-oua*, he seemed puzzled, and looked at me as if he thought I was mistaken.　Though he can quite well distinguish donkeys and horses and oxen in the street, in his wooden menagerie they are all called *moû* (ox).　The other day I had set him up on my table, with a pencil in his hand and a piece of paper before him. I drew for him roughly the outline of a quadruped, and he instantly said *moû*, thus showing that it is the dominant impressions which translate themselves, as general and individual ideas, into the words most easily and firmly retained.　Other words are useless to the child ; they do not interest him, and signify nothing to him.　If he is forced to learn them like a parrot, he forgets them more easily than those which represent something to his intelligence.

These facts show that we should be much mistaken in considering rapidity of progress in speaking—setting aside the question of the organs—as a sign of precocious intelligence ; the contrary indeed seems to me often true.　I have noticed remarkable backwardness in respect to the exercise of speech, in several children of both sexes, born of cultured parents, who expressed themselves with great ease.　The father of one of these even confided to me his great fear lest his child would be dumb or partly dumb, because at the age of thirteen months he could hardly stammer out two or three words.　This child began to speak late, and he took his time even then in learning ; but when three years old, the little man had a vocabulary as rich as it was accurate, and expressed himself with very great facility and precision.　I could quote many similar examples ; and it seems to me, therefore, that the more intelligent a child is, the less he uses words, the more necessary is it to him that words should signify something if he is to learn them, and this is why he only learns words in proportion as he gains ideas about objects.　With children of little intelligence, but who are gifted with flexible organs and with a memory in advance of their judgment, words precede ideas, and often take their place; they are retained as sounds and associations of sounds, rather than as representations of personal ideas. And thus, when the children come to understand the sense of these words so long pronounced for their sound, they find that their minds are filled with other people's ideas, whereas more intelligent children, slow in learning to speak, have got a large store of original ideas, which have come to them, not through the channel of words, but by means of direct observation and experience.

Besides the habit of parrot-like repetition, which I have explained as being caused by the fact that certain ideas and sounds take possession of children, we may also notice in the most intelligent of them a mania for jabbering strings of syllables without any meaning, jumbled together at hazard. A little girl, two years and two months old, went on repeating from morning to night for a fortnight, *toro toro, toro toro, rapapi, rapapi, rapapi*, a rhythmic monotone which caused her great delight. Another child, nearly three years old, has a stock of such refrains, which he either speaks or screams, and which he sometimes produces for fun in answer to any one speaking to him. For three months he went on repeating these three syllables, articulated in a sonorous voice, *tabillé, tabillé, tabillé*. No one understood the meaning of them, nor did they seem to have any meaning for him. His father confided to me his anxiety as to the child's intelligence. " I do not know if I am mistaken," he said, " but I am less satisfied with his intelligence than I was. Formerly he seemed to be observing and thoughtful; but perhaps this was only a passing and deceptive ray of light. I now find him heedless and incapable of attention, or of following out an idea. Sometimes I fear that his brain is softening." I succeeded in reassuring the father by giving him the following explanation of his child's jabbering propensity: The feebleness of his intellectual organs, over-exerted at the time of his first attempts at speaking, cause him to seek rest and amusement in these mechanical babblings without any sense, which he has no trouble in producing, and which do not excite his brain, and which, moreover, charm and please his ear. Savages, in like manner, will go on for several hours at a time making a monotonous melopœia, whose easy rhythm distracts their minds from care, and allows their imagination to disport itself agreeably. It is, no doubt, for a similar reason, that the dreamy and idle peasants in the South of Europe, while watching their flocks or working in their fields, will repeat either a song, or a couplet, or a refrain, a hundred times in the same afternoon. This is habitual, agreeable, and easy to them; a triple reason why they delight in the practice. There is nothing in this meaningless reiteration to cause anxiety with regard to children, when it does not fill up all their time. It merely distracts them and relieves the fatigue of their brain.

It is a universal fact that children always take hold of the superficial qualities of things rather than their substance, of the simple rather than the complex, of what is easy before what is more difficult. Thus we find them more apt at imitating the emotional tone, the emphasis, and the most sonorous syllables of words, than the entire words. I have already said, their attention and powers of enunciation are chiefly directed to accentuated syllables and terminations. "If a child is not able to reproduce a phrase it has just heard, it will at any rate reproduce the emphasis, which is, as it were, its music. If we try to teach a child the habit of expressing thanks for anything he receives, and if he cannot repeat the phrase of thanks that is taught him, he sings it without words, or at any rate without the consonants, and generally on a single vowel."[1]

Philologists who have studied children, mention a number of similar facts, all of which they refer to the general laws which govern the formation of languages. We would specially refer our readers to two philologists, already cited,

---

[1] E. Egger, op. cit.

M. Egger and M. de La Calle, in whose works they will find an abundant collection of valuable observations on the progressive evolution of infant logic and dialect, showing the gradual extension and generalization of the meaning of the words they have learnt; their first attempts in the art of forming phrases without the use of the verb *to be*, by the simple juxtaposition of a noun, a pronoun, or an attribute; the inflexible logic of analogy, which makes them confuse in their mind the thing, the person, and the action, so that they soon begin to designate them by uniform and general appellations; the incoherence and indefiniteness of their notions of time and mood, of number and of person, which makes them first use all verbs in the infinitive, and afterwards, by dint of analogy, put them in the times and persons used by those who are speaking to them, *i.e.* taking the second person for the first, and the first for the second; and so forth. On all these cases of progressive and parallel development, often contradictory in appearance, the writers above mentioned have contributed valuable information.

# CHAPTER XII.

## I.

### THE MUSICAL SENSE.

As soon as a child begins to distinguish sounds clearly, we notice that some appear to please and others to displease him ; this may arise either from the fact that particular tones correspond to certain conformations in the child's acoustic apparatus, or, by virtue of hereditary predisposition, to certain innate conditions of his personality. We have also said that children very easily accommodate themselves to sounds which are very disagreeable to adults, provided they possess some sort of rhythm, however rude. In short, we have traced in young children the awakening of a kind of musical sentiment, manifested either by their rhythmic movements and joyous expression of face when any one sings or plays on the piano, or by the evident enjoyment with which they imitate the cry of an animal or the song of a person. We must now endeavour to define these indications more precisely, and to show the genesis and the nature of the musical sense in young children.

The primary element of musical impressions lies in the emotional character of the sounds. The vibrations of matter transmit themselves to the ear under the form of sounds. In these sounds then the ear seizes certain vibrations of the bodies producing them, something, *i.e.*, of their life, so to speak. It is, first and foremost, this expressive power of the inner nature of things which awakens sympathetic sentiments in the hearer ; and with it are combined the three essential characteristics of sound—pitch, intensity, and tone, and the quality resulting from the duration of the note, rhythm, measure, and movement. In proportion as the pitch is high, a note or sound becomes more distinct, more striking, and, within certain limits, more agreeable. The medium intensity between a feeble sound which is hardly perceptible to the auditory nerves, and a violent one which shocks them, is also productive of agreeable sensations for the ear. The subtle relations of tone and feeling afford one of the most powerful means of expression. With regard to rhythm, which is both a soothing and a stirring influence, though it strikes the nerves as a shock, its periodicity and order appeal to the intelligence.

Let us examine the first manifestations of musical sentiment in children on the lines of these æsthetic principles. There is no doubt that the pleasure they experience in listening to singing or pianoforte playing springs from all the causes we have enumerated above; the human voice pleases in itself, and so in a lesser degree does the sound of an instrument, because both convey a vague expression of existence. But this expressive character of music is very confusedly apprehended by infants or animals. I doubt even whether, before the age of six weeks, a child distinguishes at all clearly the tones of caressing, of pity, or of affection, from those of calling, threatening, command, or anger; he soon arrives at understanding them, thanks to the inherited powers of adaptation in which his organization is so rich; but the differences he feels in them at first are physical rather than moral. It is necessary that associations of ideas of pleasure and well-being, of pain and uneasiness should intervene in the development of these functions, latent in the organization. By the time a child is three months old a slight advance has already been made in this direction. The child begins to distinguish with clearness and with pleasure certain sounds which to him are expressive; but he is quite indifferent to others, whether clear, harsh, or grave, which on an adult would produce either pleasant or painful impressions. The tones of authority and anger in his nurse's voice (and still more in his father) are clearly distinguished by the child from those of coaxing or fun. If light, sharp sounds enliven him, it is more on account of their pitch than their *timbre*. But sounds of all sorts, sharp or flat, intense or feeble, impressive or not, give him pleasure, provided they are accompanied by rhythm.

Of these different musical elements those which will have made most way by the age of six months, are those which appeal in some sort to physical sensibility. I mean to say, that sounds which transmit to our nervous system the very vibrations of the matter itself will be more differentiated and more keenly felt by the ear. It is the same with the sonorous vibrations which not only affect the tympanum but which shake the whole body more or less strongly.[1] Sweet and clear sounds, or harsh, jingling ones, always please children in themselves ; but they please them much more when they are rhythmic or reiterated. A child placed close to a grand organ will cry and writhe under the sonorous vibrations which shake its whole body ; the noise of a drum very near has often the same effect. But heard a little way off, the same sound, softened by distance, will please him immensely. Thus it is the striking forms of emotional music rather than its expression, the purely physical percussions and sensations of sound, and above all the physical excitement produced by the mechanical excitation of sound, that a child of six months feels the most keenly.

At this early period it would be premature to look for genuine musical intentions in the first monotonous babblings which children seem to make with no other motive than delight in the exercise of their little voices. The utmost we can admit is, that lively airs and harmonious sounds produce more marked impressions on them than do grave, sweet, and even melodious sounds ; and that their irregularly rhythmic utterances, vaguely and feebly expressed, are perhaps an attempt to reproduce those sounds which please them. There is no doubt that in addition to the immediate pleasure of exercising the vocal organs, there comes in a vague discernment of the variety of sounds, a subtle sense of pleasure in this variety, and the still more subtle pleasure

---

[1] G. Guéroult, *Du Rôle du Mouvement dans les Émotions Esthétiques. Revue Philosophique*, Juin, 1881.

of attempting to imitate sounds—his nurse's voice when singing, or the notes of an instrument.

Both language and music are in their origin but one and the same thing; they are first and foremost the expression of sentiments little or not at all defined. Thus children during the first few months practise the two simultaneously, and more for pleasure than to express any wants. Adults, however, soon teach them the separate use of these two arts which in primitive man also were fused in one, but which have been brought singly to perfection by the slow elaboration of ages. First, between the ages of one and three, children learn a very great number of oral signs and combinations of signs representing concrete ideas. But during the same period they will scarcely learn to distinguish the abstract scale of sounds better than that of colours. The child's vocal music is of a different kind, and cannot accommodate itself to ours. Not that his ear does not seize with tolerable clearness the more delicate shades of sound, provided people sing to him in tune; but his ear at this age is more natural than musical—it distinguishes musical sounds more from their accent and rhythm than their expression. A child must first have learnt by a long apprenticeship of hearing and sight to distinguish the nature and origin of any sounds whatever, to recognise the voices of different people, to imitate their speech and their songs, or to mimic the cries of animals or the most striking sounds produced by objects, before notes sung by the human voice will have any marked expression for him. And further: the development of the musical ear is quite as much aided by the personal exercise of the voice as by hearing the songs of other people. Now, the larynx of a child, though it contains at birth all the same parts as it does in an adult, is only a rough sketch of what it will be. During the first months the cartilage is in an essentially fibrous state. It is not until after the second year that the thyroid cartilage assumes more marked forms, and that the arytenoids begin to take shape. At the age of five or six years the posterior extremities of the thyroid and other parts acquire the character of true cartilage. It is natural that the development of the voice should follow that of the organs. As the fibres harden by slow degrees, the shrill screams and wailings of early days give place to more sonorous sounds ; and by the development of the muscles the vocal pipe is placed in favourable conditions for modifying the discordant sounds of the reed. . . . Towards the first year the voice begins to acquire a more decidedly sonorous quality ; speech becomes an incessant medium of gymnastics for the voice, and contributes enormously to its development.[1]

From one to three years of age the development of the voice appears to coincide with that of the brain, rather than of the general organization. With regard to its sonorous quality, it is pretty much at a standstill ; and it remains so till the age of five or six. It is from the age of four or five that the ear begins really to form, and learns most easily to distinguish sounds ;[2] and its progress has been greatly aided by the gradual development and constant exercise of the voice. A child of a year old may listen with profit to music of a simple kind, moderately expressive and, above all, lively ; it will conduce to form his musical ear. But his real musical education, that which will produce the most manifest fruits, scarcely ever begins before the age of five or six ; and even then, his best model for vocal music will not be a man, whose voice is an octave lower than his, but the sweet voice of his mother or nurse, or of an older child whose voice is already formed. As for instrumental music, we know what it is, what it may be, and also what it should not be at this age of animal *virtuosité*. We have already seen, according to Houzeau, that the first rudiments of instrumental music are not unknown to certain animals ; and that their first attempts in this line are of the nature of drum performances, that is to say, they consist in striking sonorous objects with sticks. The drum, according to Houzeau, is the universal instrument, the symbol of musical art in the savage, and, perhaps we may also add, in the quadrumana also. To this instrument of barbarous per-

---

[1] De la Calle, op. cit., p. 141.
[2] Dupaigne, *Conférences Pédagogiques*, 1878.

cussion, the offspring of the civilized bimana adds the trumpet, symbol of a more refined savageness. The use of the first of these instruments has been suppressed in France for adults, let us hope that it will not be long before children also are forbidden its use. As to the suppression of the trumpet, this last vestige of animal combativeness, it will come in its turn, though the time does not yet seem near at hand.

## II.

### THE SENSE OF MATERIAL BEAUTY.

At the end of the first month or towards the middle of the second, the fixity of expresson, the sustained attention, the smile, the automatic gestures of the head, arms, and legs, which we notice in children when they see before them brightly coloured or luminous objects, or objects moved briskly about, do not appear to signify anything more than the pleasure resulting from very exciting sensations. At this period also the sight of a candle, or anything of pronounced colour, will cause starts and tremblings and babblings, which are the child's ordinary expression of joy, admiration, or desire. For some time already the sight of his feeding-bottle, his nurse's breast, his parents and friends, will have evoked from the child analogous cries, gestures, and attitudes. During the first month, therefore, we may assume that the child confuses the *beautiful* with what he likes. The child is at the stage of the first purely animal emotions, the accumulation of which has produced the hereditary instinct called æsthetic. We are already able to affirm that the intensity of these visual pleasures is in relation to the individual impressionability, and we can perhaps also vaguely foresee the degree of the future development of this force. Psychologists, however, must observe extreme caution and reserve in their diagnosis, for these first indications have only a very limited object ; they only bring into evidence the feeblest of the elements of which the æsthetic sense will eventually be composed;

besides which, inherited tendencies, especially when pre-
cociously displayed, **are apt to** become very **mediocre** in
quality.

Let us study a child at the age of ten months. A great
number of visual perceptions have become associated in his
brain with the admiration, joy, sympathy, and desire which
the sight of **anything** good or pleasant awakens in him;
nevertheless, **in spite** of some **progress** which he has made
**in the** habits **of** imagining, **comparing**, abstracting, and
**generalizing, it seems** that the legacy **of the** ideal inherited
**from his** parents **has** not **yet** become **amplified.** The
**æsthetic** pleasure of admiration and purely **sensual** pleasures
seem still blended together. I give a **cake to a** child of
**nine** months; **he** reddens with emotion, **and his whole**
being is agitated; **he** stretches out his hands **eagerly, and**
carries the cake to his mouth with **the most unconcealed**
delight. I then present him with a plaything—his sister's
doll; **his** delight and admiration **are** shown at **first by the**
same signs as before; **but very soon** discovering that **this**
charming **object is only good to** be looked at and handled,
he confines **himself to** enjoying it **with the** two senses of
sight and touch, and presently **even** invites me to share
his pleasure. Here we have a sentiment less egoistical—
or rather an egoism which takes him out **of** himself—and
which the very nature of the object has led the child to
experience. We can see in this a progress, though very
slight, of the æsthetic **sense.**

I see a child **of ten months in a high state of delight just**
after his **nurse** has put him on his smart new frock and
**shoes.** But **I also** observed that **in** putting them on she
said, " Pretty, pretty," a word which he applies in his mind
to everything that is good or pleasant; I **also** reflect that
**every** change relating to his person **makes him** happy,
especially if the people about him appear to share his joy.
But then I remember **also that he** seems always very happy
in his everyday **frock** and **shoes, and I** can no longer
attribute this momentary excitement to a sentiment,—except
**perhaps a very** obscure one,—of the beauty of his clothes.
**The colour of the** stuff **no** doubt pleases, as that of a flower

or piece of coloured paper would; the rustling affects his tympanum pleasantly; the click-clack of the little new shoes amuses him, as any new sound would; but this is probably all.

Another child of ten months and his cousin of thirteen months can distinguish easily amongst five or six other kinds of food the cake or dainty they prefer; and if they seize hold of them it is with full consciousness of what they are doing. But if I present to them at the same time several toys and dolls of unequal prettiness, they generally make a grab at hap-hazard and without choice. If they do make a choice, it is anything but an æsthetic one; it is size, smartness, and novelty which attracts and fascinates them.

With regard to animal beauty, and the highest kind especially, that of the human form, there is a sympathy of origin and similarity, combined with the influence of familiar experiences, which predisposes children to take lively delight in it. But there is little more than this.

I have often studied little children in the presence of animals at the Jardin des Plantes in Paris, or the Zoological Gardens of other great towns, and it is evident to me that they take great notice of them and delight in looking at them—big ones and little ones, pretty or ugly, and especially those which resemble animals they are already acquainted with. But I have examined the expression of their eyes and faces, their gestures and exclamations, to learn if they made any distinction, in virtue of hereditary instinct, between the different representatives of zoological species, and I confess that, to my great astonishment, I could discover nothing of the kind. They showed quite as much delight at the tricks of the monkeys as at the gambols of the bears, or the grave and ponderous attitudes of the elephant; they admired with equal enthusiasm the brilliant cockatoos, the hideous vultures, the grotesque ostrich, and the graceful ichneumon; and would gaze with pleasure quite unmixed with horror at the awful boa-constrictor and the scaly lizards. The child has not yet got beyond isolated perceptions, and is incapable of conceptions of *wholes* or masses; and that is why his ideas of the *beautiful* and his correlative ideas of ugliness are so incomplete, variable, and fleeting.

The idea of proportion and suitability, which is a wholly intellectual perception, takes longer to form itself than the discernment of expression, which is almost entirely sensory. The attitude of these little children in the presence of people whose faces are unknown to them, seems to indicate this.    They are attracted at first sight by certain faces, which also please adults ; and other faces which do not please us seem also to frighten and repel them.    But the readiness with which they become reconciled to the latter, provided they discover in them signs of benevolence, and the equal readiness with which they withdraw their favour from the others if they only find coldness in them, authorize us in supposing that if hereditary influences and, up to a certain point, personal experience, dispose the child to feel the charm of a beautiful face, of a harmonious arrangement of form and colour, a stronger tendency makes them capable of understanding and feeling the true expression of sentiments which are not very complex.    Even with adults expression ranks before beauty of proportion.    The best proportioned face, if wanting in expression, says nothing to us, whereas the most irregular features—even the most repelling—if lighted up with expression, interest and please. It is not surprising then that to children the intellectual elements of the beautiful should be subordinated to the sensory ones, or even entirely absent.

We have now come to a fresh stage in the slow evolution of the æsthetic sense.    The child is eighteen months old ; his mind is stored with a considerable number of perceptions more or less well differentiated and generalized ; he has made, and has heard made, quantities of judgments implying a conception of the *beautiful ;* and this term, often used by him or in his hearing, may have assumed the form of an elementary abstraction.    But how undetermined still and fluctuating is this idea in his mind !    To him the beautiful still means only what is *pretty,* but it is also what is nice, and in both cases it is the concrete expression of the *known.*    All his æsthetic judgments refer always to

T

himself; *pretty* applies to everything that is part of himself, that belongs to him, that is for him or near him ; his person, his clothes, his toys, his parents, his friends, his animals, flowers, and trees. Sometimes, however, all these things cease to be pretty—himself, when he gets angry, or has been disobedient, or given pain to some one ; his toys when he is tired of them or has broken them or made them dirty ; then nothing is pretty or good. Thus we see that the dominant elements in the infant's sense of beauty are the primary judgments and sentiments, or those immediately derived from them, which make up his young personality.

A very few examples will show how very limited is that part of his ideality which may be called rational, and quasi-universal. A gaudy picture-book will drive children wild with delight, even at the age of three, while the paintings of a master do not appeal to them at all. Beautiful statuary leaves them indifferent; but they will watch with the greatest interest the tricks and antics of a dog, the flight of a bird, a boat gliding along. A little girl twenty months old, who had been accustomed to looking at pictures at home, and took pleasure in recognising the objects they represented, escaped from her father when he tried to make her look at the animals and people in the pictures of the Louvre. Her chief delight here in public was to run about, laughing and calling to me in a little shrill voice, and getting in the way of the visitors. Another child, three years old, after having looked at an Italian picture with very bright colouring, partly from imitation and partly from obedience, expressed his admiration thus : " It's very pretty, papa ! There's lots of gold, lots of red, and lots of blue ; and then, down there, there is a papa and a mamma ; there is no baby ; and there is a papa tree and a mamma duck."

Children appreciate rural scenery as little as they do pictorial or sculpturesque beauty. In Touraine, a beautiful and open country greets the eye at every turn. A young child will pass some time in the midst of its beauties, and will perceive only himself or his parents. The latter adroitly lead the cnversation to the subject of the lovely country ; the child mechaniclly repeats some fragments of their dis-

course.   Then the parents, seated on some rising ground,
invite the child to look at what they were admiring.   His
reply is very brief: "Oh! yes, it is very beautiful, very
beautiful, much more lovely than at home, and than where
grandmamma lives too."    Before a thundering cascade,
glowing with rainbow tints, a child of the same age cried
out : "Say, mamma, why is the cascade of the mill-stream
at Tarbes not bigger than this?"    Another child, about three
years of age, imitating her mother's example, always admired
the beautiful *Pic de Ger*, which towers in the distance far
above the mountains surrounding Eaux-Bonnes.    This
mountain, situated nearly to the south of the town, changes
its aspect according to the time of day.   The child had
heard this spoken of, and repeated the words in his own
fashion : "The mountain is very large!   This morning it
was quite white, yesterday it was quite black, and the other
yesterday it was all rose-colour.   Oh! the lovely mountain.
It is a great deal bigger than our house, perhaps it is four
times as big."   Of a beautiful animal, this same child said
it was of such and such a colour, and then, very large or
very good, not naughty or bad at all : of a very fine poplar,
that it was very tall or very pretty, but not so large as the
fig-tree, the big fig-tree in grandma's garden.

   Children begin by feeling pleasure and admiration for
isolated objects, and so much the more as they appear to
them to be good or pleasant.   The measure of their approbation does not go beyond their familiar experiences.   Of
masses they only perceive the general bulk ; of harmonies in
nature or in art, only the colours and the most salient points.
The ideality transmitted through ancestors, develops
according to the laws of general evolution, adapting itself
gradually to more distant objects, analysing and combining
them more and more.   The more persons and objects
recall real connections and distinct associations of agreeable
and intense sensations, the more we may say that the sense
of the intellectual element of the beautiful, or ideality, has
progressed.

## III.

### THE CONSTRUCTIVE INSTINCT.

"Creative imagination," under the form of constructive or destructive mania, "shows itself at a very tender age," says Father Girard ; "for if the little child likes to give proof of his strength by destroying, he also delights in producing, after his own fashion, things new and beautiful. See how he ranges his little soldiers, his toy horses and sheep, etc. ; how he rejoices in new combinations ! And he calls to his mother, that she too may share in his pleasure." [1] The instinct of imitation, so active in all young animals, conduces to the rapid development of this hereditary faculty. Tiedemann points out to us in his child, thirteen months old, a rare aptitude for combining the ideas he had acquired and of applying them to actual perceptions, with the evident intention of representing the first by the second.

On the 29th of October the child got hold of some cabbage stalks and amused himself with making them represent different persons visiting each other. The philosopher rightly sees in this the germ of the poetic genius, which he says, "seems to consist in transferring known images to strangers." But he forgets to tell us how far this performance was genuinely spontaneous, and whether the child had ever seen this done before, or whether it was his original idea for representing scenes which he remembered. It is true, however, that even if the action did proceed from mechanical imitation, as consciousness would soon intervene, it would assume a certain personal character. Possibly also a future *savant* was exceptionally gifted with precocious talents.

However this may be, this imaginative faculty shows itself, with greater or less force, in all children from the age of eight or ten months. A child of nine months, seated on the floor in the middle of a room, seemed like a creating and despotic diety in the midst of his playthings, and any-

---

[1] *De l'Enseignement Régulier de la Langue Maternelle*, liv. iii., p. 88.

thing else that was given him or that he could get hold of
by crawling along. Trumpets, drums, balls, paper, books,
cakes, fruit, were piled up together, ranged side by side,
separated, put back higgledy-piggledy, pushed away, fetched
back again, hugged up, kissed, gnawed, etc., etc., and all
this with an accompaniment of shouts, gestures of admiration, and bursts of joy, which showed his imperative need
of exercising his physical powers, of satisfying an ever-new
curiosity, and of imitating; and also his intellectual and
moral necessity of realising **an ideal** corresponding to his
faculties, "of producing, to the best of his ability, something
**new and** pretty." Thus the infant-man **constructs and**
destroys in play, but with a seriousness and purpose which
reminds one of Sallust's patricians, who unceasingly constructed in order to demolish, and demolished in order to
construct.

The æsthetic faculty may even at this early age receive
an appropriate kind of culture, through the development of
the constructive instinct. Give a child of twenty months or
**two years** a spade and a little pail, set him down on a sandy
**beach, and you will have reason** to admire his indefatigable
and ever-varying efforts at building and demolishing and
rebuilding. His **imagination** outstrips ours, **in that** it does
not know how to, or cannot limit itself. I saw the other
day in one of the squares which I frequently visit, a little
girl seated by her nurse's side, who, during a quarter **of an**
hour, never ceased filling her little pail and turning **out**
moulds of sand. After she had turned the pail over, before
lifting it up, she always patted it several times with her
spade; this, the nurse told me, was to make the mould
**even.** But she did not always succeed in producing a
perfect **shape,** and then she would turn to her nurse and
hold **out the** spade and pail to her, as if inviting her to share
her labours. To let children work at what they can do and
what pleases them, seems **to me** an excellent plan.

Whilst encouraging, however, we must be **on** our guard
not to over-cultivate this constructive instinct, which is as
imitative as it is inventive, and as clumsy as it is irrepressible.

One day, thinking that I should greatly interest one of my nephews, a little fellow rather more than three, and very intelligent for his age, I said to him in the garden, that we would make the river Adour, with the bridge, and the poplars on its banks. With the end of my cane I scooped out a long furrow several inches wide. I broke off some arbutus twigs and planted them on each side of the furrow to represent the poplars; and I constructed the bridge with the lid of a box supported on heaps of stones which represented the piles. When it was finished I asked my nephew if it was not very pretty. He replied : " No, it is not pretty." I was not discouraged, however; I emptied two large pails of water into the furrow, and the water trickled slowly down producing a stream which I dignified with the appellation of the Adour. Still, however, I got no admiration for my pains. I then made two paper boats and launched them on my river, which I had first replenished with another supply of water. The boy, who was very fond of boats, instantly seized one of them and put it back on the river, now almost dry again. I poured in a fresh torrent, which, being too impetuous and abundant, submerged the frail craft. The child cried out : " But there are no boats on the Adour ! They are on the Garonne, and they don't do like that on the Garonne. No, that is not amusing, no, not at all." I thought it was useless to argue the matter, and I walked off, laughing to myself over my clumsy attempt at infant construction. I had already often read in books, but this time I had learnt experimentally, that the initiative of children is always superior to what we try to devise for them. This experiment and many others have taught me that their creative or poetic force is much weaker than is commonly supposed.

## IV.

### THE DRAMATIC INSTINCT.

The imitation of gestures, sounds, and the cries of animals, also indicates in all children a first awakening of the æsthetic sense. One child of eleven months used to clasp his hands together, as he had seen other people do, to express astonishment or joy. He used to imitate all that pleased him in other people; and very often the mere fact of imitation was a pleasure to him. I had given him a little bucket, which he used to hold in his hand while walking with me in the garden. One day I threw a pebble into the bucket; instantly the child picked up another pebble and threw it in also, and then he rattled the bucket with all his might. One lesson was enough to teach him a game which he repeats frequently; it consists in covering his head with a table-napkin or shawl, and saying, "*a pu*" (*il n'y est plus*), and then popping his head out again.

His aptitude for imitation showed itself very curiously, up to a certain day he had always needed some one to help him walk, but only one person. If he was with his father, directly he saw me, he would call me *papa*, ask for my hand, and leave his father as soon as he had hold of mine. He would do the same to all new-comers—let go the hand he was holding to take theirs. One day I noticed that he kept his eyes fixed on my feet all the time we were walking, and to amuse him I began taking very long strides, which made him laugh very much and jump about with delight. I repeated this amusement the next day; the child then, to imitate me, stretched out his legs till they were almost horizontal. Naturally he lost his balance and fell down. Then he asked for the hand of the next person he saw, and without letting go of mine, he began to take these long steps again. Now, if there are two people near him, he must always have a hand from each, in order to perform this peculiar walk.

We know how early children's games take the form of dramatic scenes. When only three or four months old, their whole bodies become agitated and they utter cries of delight if their mother or sister hides for a minute behind a handkerchief or an apron. They will repeat before visitors the games which amused their parents. Almost all children display, in varying degrees, a tendency to perform monkey tricks, to say absurdities, and to repeat strange-sounding syllables, for the edification of those around them, and especially to win the admiration of strangers. I know two such children, the eldest of whom is more than three, whose mania for this kind of thing renders them almost insupportable : the presence of a visitor excites them to such a degree that they have to be sent out of the room. The eldest especially never makes a gesture or movement without turning round to the visitors to see if they are looking at him ; he seems to think they have no eyes or ears for any one but him ; and that they are only there to laugh at him and his tricks.

We must endeavour to preserve in children and their games their primitive character of innocent mischief and fun. Mockery, which is an odious tendency, borders very closely on the fun which delights us. I think that in general this fault, even when hereditary, does not appear in children under the age of three unless it has been developed by example and encouragement. The sense of the ridiculous seems to be very weak at this age. We must not imagine that because children very early show a tendency to pick out and imitate the physical defects of people, they have any notion that they are defects or absurdities. It is only that they are astonished at singular appearances and shapes and want to know the reason of them.

A child four years old, having seen in the street a man very much bent, and at another time a very little old man, observed them very attentively and afterwards asked how they were made like that. His brother, two years and a half, made remarks of the same kind: " Why does he walk like that, mamma ? " But there was no perception of the grotesque, as such, in either of the children.

A child of three years, having stayed for three weeks with some relations, brought back with him some ugly mocking habits. He had sometimes seen in his walks a little old hunchback, and he had taken to mimicking him, bending his back double and taking little hurried steps; and the servants had laughed at and encouraged this pantomime. The family too treated it as a laughing matter. From this time forth the child's chief delight was to *act the little old man*. Numbers of parents are too ready to encourage this love of caricature because they see in it the sign of a vivacious nature and an observing turn of mind. But this is one of the worst uses that intelligence can be put to. Ridiculing and caricaturing, unless for legitimate ends, are anything but pleasing in adults, and certainly not in children, who have no call to practise them. "These mocking, comedian manners," says Fénélon, "have something in them that is low taste and contrary to all good feeling."

Plenty of fun in their games, and plenty of merriment, but no mockery, is the best rule for all infant dramatizing.

This tendency to imitate and dramatize may easily have been derived from imitations which were both harmless and instructive. Before they are fifteen months old, most children will counterfeit very drolly the voices, the songs, and cries of a certain number of animals. This is a very innocent use of the comic faculty, and it has the advantage of developing the vocal organs, and of leading children to study the cries and habits of animals. It is very easy to prevent this tendency from degenerating later into an unkind, unbecoming habit.

# CHAPTER XIII.

## PERSONALITY.—REFLECTION.—MORAL SENSE.

THE idea of the *ego* is a very obscure one both for psychologists and physiologists ; for each individual, however, it is very distinct. Human beings and animals very quickly learn not to confound themselves with the people around them. This idea, moreover, is only a progressive development of the innate sense of personality which we have alluded to before. We may consider it up to a certain point as innate and hereditary, and as being already exercised and nourished by impressions during the fœtal life ; in short, as forming part of the cerebral predispositions which the child brings with him into the world. It strengthens afterwards, and becomes more defined in proportion as the organs develop, as experiences multiply and comparisons extend, and as the power of abstraction and of generalization progresses.

In the new-born child the personality is, or seems to be, especially concentrated in the emotional sphere. He does not distinctly recognise any object or himself ; but he feels the presence of objects, and he feels himself *live* and *feel* and *act*. A baby of one month has just been brought into my room. He scarcely knows his mother's voice, and he does not respond to it by any gesture, nor by smiling, which action he still performs automatically and without any meaning. But he is beginning to know his mother's breast by sight. For the last ten days he has been able to see sufficiently to distinguish objects. Before this he only followed with his eyes the movements of a candle ; now he follows all brilliant or moving objects. He has followed the movement of my finger which I waved before him a few centimetres off; he has also followed a sheet of white paper which I shook a little further off. But nearly all the time he has been in the room his eyes have been fixed on an object placed near the window, or the window itself; I think it must have been a picture-frame which glittered in the light from the window which riveted his attention. It seemed to me for a moment that he turned his head towards me, *i.e.*, to the left, to listen when I spoke; I placed myself on the other side, and spoke in a loud voice; he turned again, but slowly, perhaps because the sound came rather from behind. He performed, as if unconsciously, a balanced movement from up to down with his left arm; I touched his cheek with my finger and then with a pen-holder; he reversed the automatic movement, but his arm did not go up to his cheek. He made the same movement when I rubbed the tip of his nose gently, accompanying it with a very rapid twitch of his left nostril and a puckering of his forehead, which lasted for three or four seconds. His mother having pushed his cap a little aside, and then covered his head with a veil, he puckered his forehead, opened his mouth with a grimace, and uttered a little cry of discontent. When he has a stomach-ache, which very often happens, he is very much disturbed, and throws his arms about, as if to repulse an unknown evil. I put my finger under his, his hand trembled a little at first, and then he squeezed my finger very tightly; I pretended to withdraw it, and he held it still more firmly, and when at last I gave it up to him, he lifted it to his lips. But he only sucked it for a few seconds, very quickly realizing that it was not the teat which nourished him. Thus then a child's impressions, ideas, and actions at this age are entirely vegetative and rudimentary, and almost exclusively, though vaguely personal His only persistent and continuous idea of consciousness seems to be of himself; all else is fleeting, dimly seen, vaguely remembered, anything but distinct.

Let us now take a child of three months. What progress we now find has been made in all the functions, and in the idea of its own personality! This small infant distinguishes his mother from the other people of the house, when they are with him at the same time as his mother ; he holds out his arms to her in preference to any one else. He is not yet sufficiently advanced in comparison to distinguish a feeding-bottle full of milk from one full of water ; but when his sister tries to take away his bottle he gets angry. He knows that *coco* is a name for the bird, and he turns towards its cage when he hears the word. He bends down towards his feet, and stretches his arms towards the bottom of his frock when any one asks him where his feet or his frock are. His mother's fondling often stops his tears when he is in pain. He smiles when any one smiles at him, and strokes his mother. Here we see all the faculties,—sensibility, intelligence, and intentional motricity,—already exercising themselves with delicacy, strength, and facility, and consciously into the bargain. He distinguishes different objects from one another and also from himself, and he can distinguish the various parts of his body. Two months more, and the association of his name with the impressions connected with his personality will be a real symbol to him, distinctly representing his personality and nothing else.

The concrete notion of personality which succeeds the primitive sentiment of this personality seems to me already completely formed when the child begins to express his thoughts. This is why I think it a mistake to attempt too much to find out what he is thinking by what he says. For instance, although the contrary is generally held to be true, I cannot admit that, because children speak of themselves in the third person, therefore the notion of their personality and the term which they use to express it, are not yet completely detached from external objectivity. When the child learns to say *I* or *me*, instead of *Charles* or *Paul*, the terms *I* or *me* are not more abstract to him than the proper names which he has been taught to replace by *I* and *me*. Both the pronouns and the names equally express a very distinct and very concrete idea of individual personality. When a three-years-old child says, *I want that*, it is only a translation of *Paul wants that;* and *I*, like *Paul*, indicates neither the first nor the third person, but the person who is *himself*, his own well-known personality, which he continually feels in his emotions and actions. An abstract notion of personality does not exist in a young child's mind.

May we not also believe that animals have quite as distinct an idea of their own personality as we ourselves have? I have a cat which answers to the name of *Mimi*, which she knows very well only designates herself; not only does she come when she is called by this name, but if any one mentions the name casually in her presence, she pricks up her ears and looks up with a significant expression of countenance; and more than this, although she is never addressed as *mother*, if any one says *the mother*, she shows plainly that she has learnt,—impossible to say how,—that this word also is qualificative of her individuality.

" No one supposes," says Darwin, " that one of the lower animals reflects whence he comes or whither he goes—what is death and what is life, and so forth. But can we feel sure that an old dog with an excellent memory and some power of imagination, as shown by his dreams, never reflects on his past pleasures in the chase? And this would be a form of self-consciousness. On the other hand, as Büchner remarked, how little can the hard-worked wife of a degraded Australian savage, who hardly uses an abstract word and cannot count above four, exert her self-consciousness, or reflect on the nature of her own existence.

" That animals retain their mental individuality is unquestionable. When my voice awakened a train of old associations in the mind of the above-mentioned dog, he must have retained his mental individuality, although every atom of his brain had probably undergone change more than once during the interval of five years. This dog might have brought forward the argument lately advanced to crush all evolutionists, and said: 'I abide amid all mental moods and all material changes.' " [1]

This being conceded, we do not think we are detracting from the dignity of infant man, if, in spite of his intellectual predispositions and the incontestably superior impressions

[1] Darwin, *Descent of Man*, etc., vol. i., p. 62.

which result from education, we recognise in him no higher
notion of personality, or power of reflection on his states of
being and his operations, than what we may suppose to
exist in the superior animals when young; while, in comparison to adult and experienced animals, we think the
young human being inferior in this respect.

Between the ages of two and four, the personal sentiment
manifests itself in an extreme degree, even in the gentlest
and best-brought-up children. A child I knew who is now
a little over three, was extremely thin-skinned up to his
twenty-sixth month : he would cry and scream for the slightest fall, and thought himself done for if he got a scratch.
He has been cured through his *amour propre*. The other
day he had a heavy fall in my presence ; he picked himself up quickly, however, after first going on all fours to
make believe he had not fallen. Another time he stumbled
on the staircase, and rolled over several times, bumping
his head pretty hard. I ran up to him, but he had already
got up and was struggling his hardest not to cry. "Did
you fall ?" I said. "No, no," he answered indignantly,
" I'm not crying, I tumbled down for fun."

This personal feeling and instinct of reflection sometimes
occasioned in him, even before the age of three, *naïf* and
touching fits of inward retrospection. He had a bad cold
and head-ache one day. He had been promised that some
little girl friends should come that evening to tea with him.
In spite of his suffering he was awaiting them impatiently,
and was determined not to go to bed. It was piteous to
see him holding his little head in his hands and saying,
" I should like to die ; when one is dead, one cannot have
a headache. If I had not been born, I should not suffer.
But I don't want to die, mamma would be so sad." When
the expected playfellows arrived, he ran up to them, and
kissed them, and then brought chairs to the fire and said
to them, "Sit down and warm yourselves ; cold gives you
headache ; " then addressing the little one he liked best
among them, he said, " Perhaps you don't much want to
dance ? Have you got a headache ?" The suffering, which
had thrown him back on himself in so touching a manner,

had opened up in his heart a well of sympathy, which quality is not so rare as one would suppose in these shallow and egoistic little beings.

The sense of personality is so strong in young children that it is always on the verge of excess. It amounts in them to egoism, self-sufficiency, and assertiveness, though they now and then have sudden rebounds of sympathy, tenderness, credulity, and self-distrust. It is from this personal feeling, the natural tendencies of which have both to be encouraged and checked, that the moral sense springs, or, in so far as it is hereditary, that it draws material for its development.

## II.

### THE MORAL SENSE.

The wholly objective notion of good and evil,—the intellectual germ of the moral sense,—cannot be determined before the age of six or seven months. I have seen a child seven months old, whose mother had taught by scolding and shaking him, that he must not scream to be taken out of bed or nursed, if his gestures and articulations to this end were not instantly attended to. But how many children are there not at this age, and even later, who, the more one tries to exact silence from them, the less they will obey? A slight thwarting of their wills results often in tremendous scenes of passion, tears, sobs, and screams. A child of ten months (and many well-trained children arrive at this earlier) knew very well how to make himself understood when he was afraid of making a mess in his cradle. He knew too that when he wanted anything which was refused him, it was naughty to persist in wanting it, and especially to cry in order to get it. He had a habit of scratching the face of people who took him in their arms, half in play and half from unconscious naughtiness; but he had at last been taught obedience in this matter. In all this we see the germ of the moral sense.

The first dawning of the moral sense appears in the child as soon as he understands the signification of certain intonations of the voice, of certain attitudes, of a certain expression of countenance, intended to reprimand him for what he has done or to warn him against something he was on the point of doing. This penal and remunerative sanction gives rise by degrees to a clear distinction of concrete good and evil. As soon as the child begins to obey from fear or from habit, he enters on the possession of the moral sense; as soon as he obeys in order to be rewarded or praised, or to give pleasure, he has advanced further in this possession.

The moral sense, in its objective form, is still very incomplete in little children, even between the ages of two and four. They have, however, a very advanced idea of what is allowed and what forbidden, of what they must or may, and what they must not do, as regards their physical and moral habits. Moral law is for them embodied in their parents, in the mother especially, even during their absence.

Last summer we had a little nephew staying with us for two months. Just at first he was very reserved and very good, as if he were on the defensive, or as if he were taking stock of his surroundings. He behaved well at table, asked politely for what he wanted, and when he had had enough, said, "I have finished." But when the meal was ended, excited by a certain amount of indulgence which he had received (unconsciously on our part), but which he had fully appreciated, he said : "At home, when I have done eating, mamma lets me leave the table. We hastened to gratify this desire, which seemed very legitimate. Then we began to play with him, and succeeded in amusing him for half an hour. At last he grew tired and burst into tears, saying: "Mamma puts me to bed when I cry." He was then taken to the bedroom and undressed. Then he began to cry again and to gesticulate in a desperate manner; he talked about his mother, and we thought he was fretting for her; but he hastened to explain that his mother did this, and that, and the other when she put him to bed; and then followed a string of prescriptions, relative to his going to bed, which he evidently regarded as so many obligations we were bound to fulfil. Thus, whatever was ordinarily done for him was his idea of what was right to do. It was the same with his own actions. He said to us : "It is very wicked to tell stories or to disobey ; it gives mamma pain, and makes her cry."

Thus, for the young child, as well as for animals, *good* is what is permitted; *evil*, what is forbidden. And hence arises frequent uncertainty with regard to new actions or to habitual ones under new circumstances. At the end of several days, the good habits of the child became modified according to his new impressions and according to the character of the persons who surrounded him. At first I inspired him with great respect. It was I who brought him to the house; and no doubt I reminded him of his father, whom he feared more than his mother. He always looked at me before doing anything that he knew was not right : he was more at his ease and less circumspect with other members of my family, who inspired him with less awe. He acquired some habits contrary to those he had formed at home. He was sometimes disobedient, and he got into the way of crying for what he wanted. I found myself henceforth compelled to make a departure from one of my principles regarding infant education, which is, to turn children's attention as much as possible away from their illegitimate desires or their little sorrows arising from trifles, and to reprimand them as little as possible and only in extreme cases. I was now obliged to raise my voice very often, and to make a gesture which he recognised as a dreaded gesture of his father's, but which I did not go the length of putting into effect.

The preceding observations show that the morality learnt by the young child is an edifice built up at the cost of great labour, patience, and prudence, and which may crumble away in a few weeks in the midst of different circumstances and surroundings.

The moral sense, then, is one of the hereditary faculties most liable to be modified by circumstances. It is not in vain that we appeal to this social instinct, this individual impressionability and sympathy, this innate desire to please, which plays such an important part in the culture of the young moral faculties. A little girl fourteen months old was always very unhappy when her mother said, "I am angry,

U

baby." But she was indifferent to most of the scoldings of her father, whom she was accustomed to hear rage and threaten. When two years and five months old, the son of Tiedemann, if he thought he had done something very good, used to cry out : "People will say, That is a very good boy!" When he was naughty, if any one said to him, "The neighbour will see it," he desisted immediately. There are delicacies of sensibility innate in children which are never appealed to in vain by good educators, and which even with children who are the least gifted in this respect are the best auxiliaries and the most efficacious substitutes for coercive means. I saw a woman yesterday passing in the street who had been shopping with her little girl about three years old. The child was walking a step or two in advance of her mother, and kept on turning round to catch a look or gesture of approval ; she had been charged with the execution of a task which she considered very serious, and she thought herself very meritorious for performing it well. She was carrying on the palms of both her hands a parcel of very large circumference but very thin and light, and which must have seemed exceedingly large to her in comparison with her own small dimensions ; and to carry it without stumbling or letting it fall seemed to her a great feat of skill. She was taking all this trouble to please her mother, and to win her admiration and praise. Passive obedience, and the fear of being punished or scolded, would scarcely have obtained a result like this, which is so easily and surely gained by an appeal to the benevolent and generous sentiments of our nature.

The sentiment of justice sometimes manifests itself with great force in young children, especially at the period when they have become able to express distinctly what they feel. The first time that a certain child of my acquaintance, now four years old, told a deliberate falsehood, his mother thought it her duty to punish him. She told him that she was going to shut him up in the cellar, and she made him go with her down the stairs which led to it. On the way down, the child, whose imagination was struck with the importance attached to his fault, and who had begun to feel very guilty, said to his mother : " But, mamma, perhaps I am not sufficiently punished for such a great fault?"

Some months afterwards, he was sent to his grandmother's, where he used to behave as a little despot and tyrant. One day, when he had perpetrated some mischief or other which was considered very serious, his grandmother shut him up in a dark room close to the kitchen. Five minutes had hardly passed when, notwithstanding her deafness, the old lady heard shrill screams like those of a peacock. She hurried to open the door of the prison, and the child rushed out, his face bathed in tears and convulsed with fear and anger. As he was only half-afraid of his grandmother, who was much too indulgent to him; he thought she had not the right to punish him so severely, and he protested against what he considered an injustice : "Oh! naughty grand-mamma, you are very naughty! I said I would never go back there, and you have shut me in there again! You want the rats to eat me! They would first of all have eaten my feet, and then I should have died. Mamma would have been very unhappy : she would never have come back here, nor my father either. I will tell him about it." It is noteworthy that children at first only apply the idea of justice to other people's actions, according to the nature of the sentiments which these actions awaken in themselves. They are irritated when any one takes away their toys ; and it is by dint of experiencing this annoyance and also from hearing it said that it is naughty to take other people's things, that they end by forming, first a concrete notion and then a tolerably general one, of illicit appropriation. Again, if one of his brothers or sisters is punished, he will come and tell you all about the punishment, the fault which was committed, and above all the manner in which the culprit bore the punishment ; and he will not fail to qualify the faulty action by some general epithet : and all this because he himself has committed like faults and consequently undergone like punishments. Children, moreover, detest injustice, especially when perpetrated, or supposed to be perpetrated, against themselves; but they see nothing more in it than disagreement between the habitual and the accidental way of treating them.

It is the same with all other moral habits, and also with the moral sense, which in children—indeed also in adults— is generally nothing more than the theory of their thoughts and actions. "This psychological law should never be lost sight of, of so great consequence is it. All our tendencies, innate or acquired, good or bad, and consequently all our habits, in whatever way we have contracted them, not only tend to determine our conduct as a matter of fact and as actual motives, but also to impose themselves as reasons to the deliberative will, or at any rate to suggest justifying sophisms, of which the corrupted judgment finally becomes a dupe almost in good faith."[1] Actions influence thoughts, and thoughts influence actions. Hence interest, self-love, and passion have their good share in the best resolutions of children. They seek to determine the motives of their actions, even when they are not asked for them. "I did this or that because . . ." This formula is constantly on their lips. They try to obtain our praises for numbers of insignificant actions which they think meritorious. They imagine motives, sometimes utterly improbable, for the actions of others, which they judge according to their own ; but above all they imagine specious reasons for explaining their own most blamable actions. What is true of habits, is also true of the moral sense ; if in one way they tend to disinterestedness, they are always "interested in some way or other."[2]

If then we wish to understand the meaning of the actions of little children, and to direct their wills in a useful and progressive manner, we must bear in mind that all their tendencies, whatever they may be, begin and end with egoism. We must especially beware of attributing to the moral sense, and with children still less than with adults, a power of determination which it does not possess. If man is far from being able to do what he wills, *à fortiori* is he far from being able to will that of which he may have a very distinct idea.

The moral sense simply furnishes the more or less con-

[1] H. Marion, *La Solidarité Morale*, p. 109.
[2] V. M. Guyau, *La Morale d'Epicure*, p. 220.

scious will with motives more or less strong according to
individuals, places, times, and situations. The business of
psychlogical educators is much more concerned with the
habits that children may acquire, and with their wills, which
are also developed by habitual practice, than with the development of their moral conscience. The latter is the
blossom which will be followed by fruit; but the former are
the roots and branches.

With the development of intelligence and thought, the
primitive distinction of good and evil enlarges and becomes
better defined. Submission to authority, and docility, also
increase with experience and reflection. With the tendency to reflection there spring up also, little by little, sympathy, the love of praise, *amour-propre*, the susceptibility to
example, the fear of reproach, the desire not to be scolded
and to give pleasure to those belonging to one. And thus
that state of mind is constituted in which reflection dominates
the contrary tendency and renders possible the government
of self. Out of all these diverse elements the moral conscience is gradually formed, which means to say that good
intention and the moral sentiment are added to the moral
habits at first acquired. A child may have been trained
more or less easily or completely according to the innate
dispositions of its character; but at this point it is still incapable of managing itself. It begins, however, to reconsider
its impulsive tendencies, it has a vague sentiment of its still
very limited liberty, and it can co-operate in the guidance
of its small personality.

What touching examples we see even in little children
barely two years old, of self-introspection accompanied by
well-grounded mistrust of themselves, and the moral pain
of *meliora video, deteriora sequor!* The daughter of L. Terri,
while still at the age of thorough impulsiveness, when en-
treated by her mother to be good, answered more than once:
"I feel that I cannot be good," thus giving proof of self-
introspection with a sense of her weakness. "At five years
old, having one day been praised by her mother, she said:
'To-morrow I should like to please you still more. I
should like to be always good. But tell me why I can't

be always good.'"[1]   It must not be supposed that even at
five years old the most gifted and best-brought-up children
will always be able to keep at this relatively high level of
moral sentiment.   Even considerably later on, they will
often fall back below what they may have been or seemed
to be at the early age of three or even two years.   The
important point is, that at so tender an age it should be
possible to awaken at intervals the ever fitful light of moral
reason.[2]

---

[1] *Osservazioni sopra una Bambina,* in *La Filosofia delle Scuole
Italiane,* Oct. 1881.

[2] For further development of this subject see in my book *L'Education
dès le Berceau,* chap. vii., on the "Moral Habits and the Moral Sense,"
from page 265 to 300.

THE END.

www.ingramcontent.com/pod-product-compliance
Lightning Source LLC
Chambersburg PA
CBHW020457270326
41926CB00008B/640